MOLECULAR FLOW IN VESSELS

RASCHET VZAIMODEISTVIYA MOLEKULYARNYKH POTOKOV S OGRAZHDAYUSHCHIMI IKH SOSUDAMI

РАСЧЕТ ВЗАИМОДЕЙСТВИЯ МОЛЕКУЛЯРНЫХ ПОТОКОВ С ОГРАЖДАЮЩИМИ ИХ СОСУДАМИ

MOLECULAR FLOW IN VESSELS

Yurii Naumovich Lyubitov
Institute of Crystallography
Academy of Sciences of the USSR, Moscow

Revised and Enlarged Edition

Translated from Russian by

Wendell H. Furry
Harvard University
Cambridge, Massachusetts

and

James S. Wood
Los Angeles, California

Springer Science+Business Media, LLC
1967

Yurii Naumovich Lyubitov, senior scientist, Institute of Crystallography of the Academy of Sciences of the USSR, was born in 1932. He was graduated from the Physicochemical Department of the Moscow Steel Institute in 1956 and worked at the A. A. Baikov Metallurgy Institute of the Academy of Sciences of the USSR until 1964.

ISBN 978-1-4899-4718-5 ISBN 978-1-4899-4716-1 (eBook)
DOI 10.1007/978-1-4899-4716-1

The Russian text, originally published for the State Committee for Ferrous and Nonferrous Metallurgy of the State Planning Commission of the USSR and the A. A. Baikov Metallurgy Institute by Nauka Press in Moscow in 1964, has been extensively revised and enlarged for this edition by the author.

Ю. Н. Любитов
Расчет взаимодействия молекулярных потоков с ограждающими их сосудами

Library of Congress Catalog Card Number 65-20215

FOREWORD

The development of many branches of modern technology rests on the availability to engineers and technicians of information relating to the properties of rarefied vapor phases above various types of condensed systems. Information of this sort is also essential for assessing the durability of components that function under high vacuum conditions and for a number of technological operations involving highly rarefied media.

Of the many possible kinds of physicochemical interaction between molecular flows and the surfaces containing them, so far only the diffusive emission of particles, without regard for conversions of any kind, has been accounted for. The theory of that type of interaction was treated definitively and completely in the thirties by P. Clausing in an extensive series of monumental investigations; a sizable portion of the present book is devoted to a survey of this work.

In view of the fact that the interaction of molecular flows with surfaces is described in some instances by laws that bear a formal resemblance to the laws of photometry, a number of the solutions obtained herein may be applied to problems in optical pyrometry, and, conversely, the solutions of photometric problems may be used for the investigation of molecular flows. The analysis is given almost throughout from the postulates of geometric optics when, for example, the laws of photometry fail to take into account dependences on the spectral composition of the beams, while the molecular flows themselves are assumed to be monoenergetic. Geometric optics overlooks such an important property as transparency or transmissivity. This bears upon the relatively low penetrability of molecules through solids and liquids.

The present book mainly examines some of the results of the theory of molecular flows through cylindrical containers of finite dimensions, taking into account many more kinds of interaction between the flow and the surfaces that contain it than does Clausing's theory.* In addition, results are given for the spatial distribution of molecular flows after emergence into the open from a cylinder, for the motion of molecular flows through short conical and circular bent tubes, and for the motion and chemical interaction of molecular flows in infinite cylindrical tubes, the more general aspects of molecular flows in vessels are discussed in the Appendix.

Knudsen molecular flows are normally assumed herein, but in order to complete the picture, some of the possible transitions to intermediate and even Poiseuille flow are also indicated. It is the author's hope that the book, a certain synoptic slant notwithstanding, will be helpful in comprehending the wealth of material on the theory of molecular flows in vessels.

Such a comprehension is mandatory both for engineers and for scientists, because an analysis of the literature reveals, first, a recurrence of the same solutions (not always successful ones) and, second, a frequent disinclination in original works to enumerate the restrictions and assumptions by which the problem is idealized, leading to a lack of understanding as to the limits of applicability of the final solutions.

We have not endeavored to provide an exhaustive answer to all the problems that are discussed, posed, or set up for formulation. Very often certain aspects are considered in less detail than in the original works. The aim of the present book is to generalize present-day methods of solving the problems stated.

The author gratefully acknowledges those who have contributed to the writing of this work. They include the reviewers, A. M. Yakobson, O. N. Talenskii, and E. Z. Vintaikin; fellow-workers E. S. Kuranskii, D. V. Kor-

*The works of A. I. Ivanovskii and A. I. Repnev on a series of related problems, published in the Transactions of the Central Aerological Observatory, are unfortunately not reflected in the present book; they are cited, however, in the references at the end of the book.

militsyn, Yu. M. Ivanov, and L. M. Babenkov; É. P. Borisov. who gave me my first lessons on the operation of computers; and V. I. Lozgachev, who was kind enough to explain some of the hypotheses that he developed. Appreciation is also expressed for critical comments offered by Profs. D. A. Kuznetsov, V. L. Tal'roze, and D. A. Frank-Kamenetskii.

The author gratefully acknowledges the moral and intellectual support of Profs. P. K. Oshchepkov and R. A. Sapozhnikov.

The work would have been delayed far longer toward completion had it not been for the unflagging assistance of M. L. Darashkevich and, especially, G. D. Kuznetsova.

CONTENTS

CONTENTS

SECTION 1

DISTRIBUTION IN SPACE OF THE MOLECULES LEAVING A PLANE SURFACE. MASS BALANCE AT THE SURFACE

Before entering the vapor phase, a molecule lies for a certain time at the surface of the solid or liquid substance, from which it can subsequently fly off in a direction which depends on many factors: the chemical nature of the molecules and of the surface, the energy state of the flow and of the surface, and the geometry of the surface.

Departing from the surface are flows of evaporating or desorbing particles, as well as flows of particles reflected from the surface.

We deal with reflection when the parameters of the molecular flow after collision with the surface depend on the parameters before collision. The characteristic quantities which describe a flow are the velocity distribution of the particles, the composition of the flow and its intensity, the energy states of the particles in the flow (excitations of various types), and so on.

A quantity which provides an indication of the extent to which a particle has forgotten its previous history is τ, the lifetime of the particle on the surface. It is found from both experiment and theory (Fraser, 1931) that τ varies over wide ranges and is a function of the temperature, given by the Frenkel relation

$$\tau = \tau_0 e^{\frac{Q}{RT}}, \tag{1.1}$$

where τ_0 is the vibrational period of an adsorbed molecule in the direction perpendicular to the surface and Q is the heat of adsorption.

As tentative values for many substances we have $\tau_0 \approx 10^{-13} - 10^{-14}$ sec, $Q \approx 100$ kcal/mole. The lifetime τ and the condensation coefficient α (to be defined later) are connected by a statistical relation (Fowler, 1935, quoted in Massey and Burhop, 1956):

$$\alpha \cdot \tau = \frac{h}{k_B T} \sum_s e^{-\frac{\vartheta_s}{k_B T}}, \tag{1.2}$$

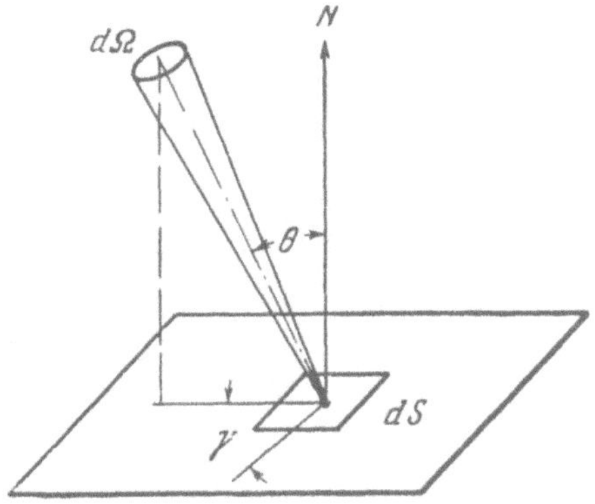

Fig. 1. Coordinate system for flow of particles emerging from surface.

where the sum is taken over all bound states of the molecules with binding energies ϑ_s, h is Planck's constant, and k_B is Boltzmann's constant. Equation (1.2) is valid only for van der Waals forces and motion of the particles perpendicular to the surface.

The average lifetime τ of a particle on a surface and the average collision period τ' are connected by the formula:

$$\tau = (1 - \rho)\tau', \tag{1.3}$$

where ρ is the reflection coefficient of the particles – the fraction of particles which have a lifetime on the surface which is much smaller than τ. For methods for the experimental determination of τ, and for calculations, see Fraser

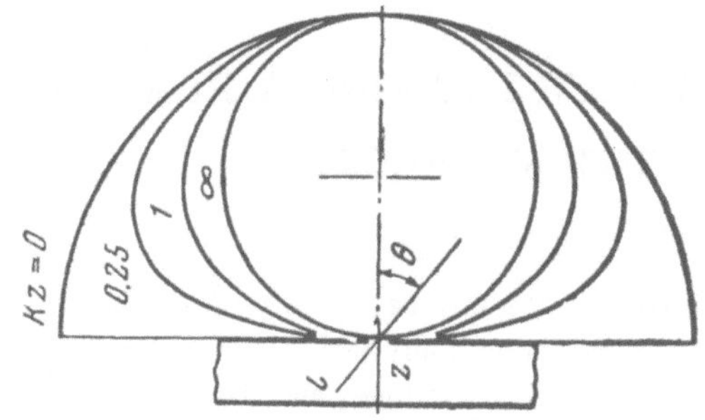

Fig. 2. Gradual transition from Lambert's law to Euler's law (Gershun).

(1931); Clausing (1930) (7);* and de Boer (1962). For mercury on glass at room temperature, $\tau' \approx 10^{-5}$ sec (Wertenstein, 1923, quoted by Fraser, 1931).

If the surface were ideally plane, a molecular flow might possibly obey the law of specular reflection which is valid for light. Actual surfaces, however, have more or less developed microstructures which cause scattering of the molecular flow according to various laws. Among the possible laws of the distribution of molecular flows in space, the simplest in form is the cosine law. According to this law, the number of particles passing per unit time through a surface element dS (Fig. 1) in a direction lying within the solid angle $d\Omega$ and forming an angle θ with the normal is

$$I = I_0 \cos\theta \, d\Omega dS, \tag{1.4}$$

where I_0 is the number of molecules which move out along the normal to dS in the solid angle Ω.

If we consider any molecule in the volume dV, it is obvious that it can be moving in any direction with equal probability. This assertion, together with Eq. (1.4), describes the molecular chaos which is characteristic of the equilibrium state. Equation (1.4) is valid for any actual or hypothetical area inside a volume containing a gas at equilibrium. A law of the same form as Eq. (1.4) is also valid for equilibrium radiation (Sapozhnikov, 1960).

In photometry the law (1.4) is called Lambert's law. In the kinetic theory of gases it was used by Maxwell, and was finally confirmed by the work of Knudsen.

Under equilibrium conditions the law (1.4) is valid both for flows moving away from the surface and for those arriving at it. In the absence of equilibrium this law has been proved only approximately and only in special cases. For evaporation processes it is usually assumed that a surface which consists of a set of dissimilarly oriented portions with projections and depressions will on the average emit molecules according to the cosine law. It can easily be shown that besides the cosine law there are other possible laws for scattering of a substance by a surface. Lambert's law, when carried over without alteration from optics into the kinetic theory of gases [M. Knudsen, 1909 (1), 1911 (11), and 1915 (15)] is the simplest law, but is by no means always correct even in optics.

For example, there is Euler's law (Sapozhnikov, 1960, p. 46), for which the intensity of light from a surface source is the same in all directions.

In a paper by A. A. Gershun in 1933 it was shown in detail how Lambert's law goes over into Euler's law. Without giving complete derivations of all the relations, we shall demonstrate the transition graphically.

Figure 2 shows polar curves of the light intensity of a point source which is on the surface of a plane infinite layer of thickness z. K is the volume absorption coefficient of the light. If the mean free path of a particle or of light is λ, then $K = 1/\lambda$. As can be seen from Fig. 2, for $Kz = 0$ we have Euler's law (law of isotropic emission) and for $Kz = \infty$ we have Lambert's law (cosine-law radiator). These laws fully retain their meaning if we apply them to the kinetic theory of gases.

*In references to sources we give first the name of the author, then the year, and, in the case of several publications, the order number in the [alphabetical] Literature Cited.

Actual emission does not ordinarily completely obey the laws indicated, since the surface from which evaporation or emission of light occurs is a set of areas of various sizes and orientations which form a rather complicated structure. As early as the eighteenth century P. Bouguer did much fruitful work on this problem, and introduced the so-called numerator of asperities (Bouguer, 1950) the vector $f(\theta)$ for which is given by

$$f(\theta) = a\cos\theta + b\cos^j\theta, \tag{1.5}$$

where θ is the angle between the given direction and the normal to the surface and a, b, j are constants which characterize the surface. Bouguer indicated a method for determining the constants a, b, j experimentally. A critical examination of the physical and mathematical foundations of this representation is given in the commentary on Bouguer's work produced by A. A. Gershun (Bouguer, 1950). Bouguer's numerator of asperities can evidently be used both to calculate the intensity of radiation (evaporation) from a surface and to calculate the scattering by the surface of external light or particles in the various directions. We give an example of the wide range over which the distribution in direction of the intensity of molecular flow can vary, depending on the structure of the surface and the type of emission of the particles (for more details see Troitskii, 1961; Naumov, 1963).

Let us consider the variation of the polar curve of emission from a many-channel source, according to the work of Naumov in 1963 (Fig. 3). Because the particle size is usually of a different order of magnitude from the mean free path it is evident that the size of the actual depression on the surface will not affect the shape of the curve.

It is a very strained interpretation to regard actual surfaces as planes with various sets of perforations, but this approximation may be regarded as permissible to illustrate the fact that the cosine law is not a unique one. It can be seen that there is a rather wide range of possible radiation curves, depending on the structure of the surface and the type of effusion of gas from the surface. In Fig. 3, L is the length of the capillary, and 2r is its diameter. The rays are labeled with the angles defining the direction of emission. Whereas for curve 6 ($2r/L = 0.02$) there is considerable beam formation, for curve 1 ($2r/L = 100$) the cosine law is rather accurately obeyed. The dependences found by Naumov are to be regarded as approximate, since the detector which he used for the molecular beam had an area considerably larger than that of the source.

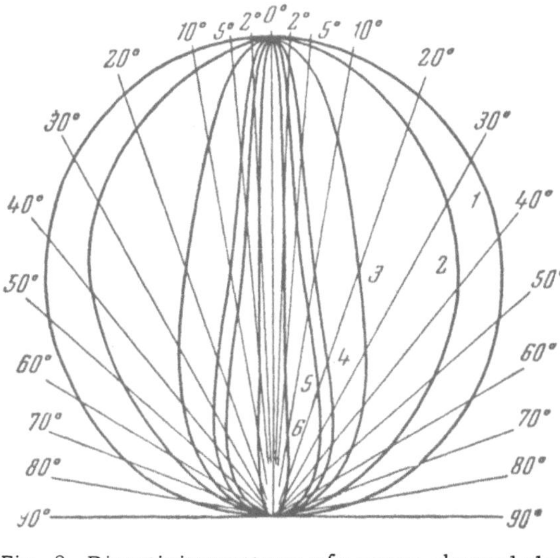

Fig. 3. Directivity patterns of a many-channel plane source (Naumov)
1 — $2r/L = 100$; 2 — 1; 3 — 0.35; 4 — 0.175; 5 — 0.115; 6 — 0,02

To obtain these curves, Naumov used sources with various numbers of channels; the curve numbered one was obtained with one channel, the numbers of channels for curves 2, 3, 5, 6 are not indicated but were evidently larger than 1, and curve number 4 was obtained with 26 channels.

In the interaction of a molecular beam with a sufficiently smooth surface, or with cleavage planes of the crystals of a number of salts, one can obtain specular reflection and diffraction phenomena. These effects are extremely slight. To find the conditions under which these effects could be detected, we refer to the notion of de Broglie waves. We compare the de Broglie wavelength with the size of the microscopic surface irregularities to determine the possibilities for specular reflection, and with the interatomic distances in the lattice to test the possibility of diffraction of a molecular beam (for more details see Fraser, 1931; Massey and Burhop, 1956; Smith, 1955; Ramsey, 1956). The wavelength introduced by de Broglie (1924), λ_B, is defined by the formula

$$\lambda_B = h/mu, \tag{1.6}$$

where m is the mass and u is the velocity of the particle. A condition for specular reflection is the equation

$$B \sin \varphi \approx \lambda_B, \tag{1.7}$$

where φ is the glancing angle, equal to $\pi/2-\theta$, and B is the average height of the surface irregularities. According to Fraser (1931), $B \approx 10^{-5}$ cm, and for hydrogen at room temperature λ_B is about 10^{-8} cm. Using Eq. (1.7) we easily find that φ must be no larger than 10^{-3}, or equal to a few minutes.

In a paper by Knauer and Stern (1929, quoted by Fraser, 1931) the dependence of the fraction of reflected H_2 particles on the glancing angle φ is given for polished bronze:

φ	10^{-3}	$1.5 \cdot 10^{-3}$	$2 \cdot 10^{-3}$	$2.25 \cdot 10^{-3}$
% reflected particles H_2	5	3	1.5	0.75

It is seen that the reflection coefficient ρ is a function of the glancing angle. Generally speaking, ρ also depends on the speed of the particles (Estermann and Stern, 1930).

The de Broglie wavelengths of light atoms are of the same order of magnitude as the wavelengths of X rays and as the distance between atoms in the lattice. In fact, in a large number of studies (Ramsey, 1956), diffraction phenomena have been detected for various molecular beams at the surfaces of a wide variety of crystals.

Clausing [1930 (5)] has shown that the use of a hypothetical law of scattering of an equilibrium gas at the surface of a closed reservoir in the form

$$I = I_0 \cdot (\cos \theta)^i \, d\Omega \tag{1.8}$$

with $0 < i < \infty$, but $i \neq 1$ would result in perpetual motion, so that it has to be assumed that $i = 1$. This proof is correct, of course, if we assume that the law is of the form (1.8). There are experimental indications (Wood, 1915, 1916) that the expression (1.8) with $i = 1$ will approximate actual conditions correct to 10^{-5}.

When the principle of detailed balance and the second law of thermodynamics are taken into account, evaporation, condensation, and the reflection of molecules under equilibrium conditions are subject to a number of laws analogous to the laws of photometry. The fact that Lambert's law and Knudsen's law are identical has already been pointed out; we now give other results [Clausing, 1930 (5)]. In Fig. 1, let there be incident on a unit area in unit time in the direction θ, γ (γ is the azimuth) a number of molecules

$$I(\theta, \gamma, u) \sin \theta \, d\gamma \, d\theta \, du = I(\theta, \gamma, u) \, d\Omega \, du$$

with speeds between u and u + du. Of these molecules a fraction $\rho(\theta, \gamma, u)$ is reflected, and the number $A(\theta, \gamma, u) d\Omega du$ is adsorbed (or condensed); we then get

$$A = (1 - \rho) I; \tag{1.9}$$

$(1-\rho) = \alpha$ is called the condensation coefficient.

We shall use primed letters to refer to a particular direction (the other variables are running variables). We can then write

$$\rho I \, d\Omega \, du = d\Omega \, du \iiint r(\theta, \gamma, u, \theta', \gamma', u') \, d\Omega' \, du' \tag{1.10}$$

and determine the number R_0 of molecules reflected per unit area and unit time in the direction θ, γ' with a speed u' from the expression

$$R_0(\theta', \gamma', u')\,d\Omega'\,du' = d\Omega'\,du' \iiint r\,d\Omega\,du. \tag{1.11}$$

For the emission E we can write the relation

$$\iiint E(\theta, \gamma, u)\,d\Omega\,du = \iiint A(\theta, \gamma, u)\,d\Omega\,du. \tag{1.12}$$

We write down the state of equilibrium chaos of the gas

$$I = \psi(u)\cos\theta. \tag{1.13}$$

The Maxwell velocity distribution function f for the number of molecules ndV which are present in the volume dV (n is the density of the gas) is given by

$$f(u)\,du\,dV = 4\pi n\left(\frac{m}{2\pi k_B T}\right)^{3/2} u^2 e^{-\frac{mu^2}{2k_B T}}\,du\,dV. \tag{1.14}$$

The function $\psi(u)$ is defined by the formula

$$\psi(u) = \frac{u \cdot f(u)}{4\pi}; \tag{1.15}$$

m, k_B and T are the molecular mass, the Boltzmann constant, and the absolute temperature, respectively

Along with Eq. (1.13) the second law of thermodynamics requires that

$$R_0 + E = I. \tag{1.16}$$

According to the reciprocity principle, which says that for each geometrical direction we will have corresponding to an equilibrium elementary act another elementary act in the opposite direction, we can write

$$E = A. \tag{1.17}$$

Besides this, we must have the condition

$$r(\theta', \gamma', u', \theta, \gamma, u) = r(\theta, \gamma, u, \theta', \gamma', u'). \tag{1.18}$$

Equation (1.17) is the analog of Kirchhoff's law. When only specular reflection occurs, Eq. (1.18) is simplified and becomes

$$\rho(\theta, \gamma, u)\,I(\theta, \gamma, u)\,d\Omega\,du = R(\theta, \gamma + \pi, u)\,d\Omega\,du. \tag{1.19}$$

A condition required for molecular chaos is

$$I(\theta, \gamma, u) = I(\theta, \gamma + \pi, u), \tag{1.20}$$

and the principle of detailed balance requires that

$$\rho(\theta, \gamma, u) = \rho(\theta, \gamma + \pi, u). \tag{1.21}$$

Substitution of the expressions (1.20) and (1.21) in Eq. (1.19) gives

$$\rho I = R_0. \tag{1.22}$$

From Eqs. (1.9) and (1.22) it follows that

$$I = R_0 + A, \tag{1.23}$$

whereas the second law of thermodynamics requires that we have the condition [cf. Eq. (1.16)]

$$R_0 + E = I.$$

From Eqs. (1.16) and (1.23) we again get the analog of Kirchhoff's law, Eq. (1.17):

$$E = A,$$

from which it follows in particular that α can be called both the condensation coefficient and the evaporation coefficient.

In many cases it is appropriate to use the Knudsen cosine law as a first approximation. This law is the basis of the majority of works on the subject, including the present one.

SECTION 2

ENERGY EXCHANGE AT A SURFACE

Let us consider the energetic aspect of the process of interaction between a molecular flow and a surface (for more details see Herzfeld, 1935; Massey and Burhop, 1956; Devienne, 1962).

This problem may be related to the effects of a vessel on a molecular beam. For example, it is easy to show that if the particles were to collide with the walls of a tubular channel in a perfectly elastic manner (according to the law of specular reflections) the tube would offer no resistance to the flow; this, however, is in contradiction with experiment.

We first derive the formula for the rate of evaporation (or desorption), a. The only monatomic particles that leave a solid body are those whose kinetic energy is sufficient to overcome the force of attraction of the remaining particles which make up the surface. Let Q_0 be the heat of sublimation per mole of the substance at temperature absolute zero, and let N be Avogadro's number. The condition for particles with velocity component u_x (x axis perpendicular to the surface) and mass m to be evaporated can be written in the form

$$\frac{mu_x^2}{2} > \frac{Q_0}{N}. \tag{2.1}$$

The fraction of the molecules whose velocity component lies between u_x and $u_r + du_x$ is

$$\sqrt{\frac{m}{2\pi k_B T}}\, e^{-\frac{mu_x^2}{2k_B T}}\, du_x.$$

Let V be the volume of a gram mole of the gas; then the density of this type of molecule is

$$\frac{N}{V}\sqrt{\frac{m}{2\pi k_B T}}\, e^{-\frac{mu_x^2}{2k_B T}}\, du_x.$$

The number of molecules striking per unit area per unit time will be

$$u_x \frac{N}{V}\sqrt{\frac{m}{2\pi k_B T}}\, e^{-\frac{mu_x^2}{2k_B T}}\, du_x.$$

According to Eq. (2.1), the rate of evaporation a' in mole/cm^2 sec is given by the formula

$$a' = \frac{N}{V} \int_{\sqrt{\frac{2Q_0}{M}}}^{\infty} u_x \sqrt{\frac{m}{2\pi k_B T}}\, e^{-\frac{mu_x^2}{2k_B T}}\, du_x = \frac{N}{V} \sqrt{\frac{k_B T}{2\pi m}}\, e^{-\frac{Q_0}{k_B T}}. \tag{2.2}$$

TABLE 2.1. Probability of Condensation in a Single Collision of Molecules with Masses m = 2, 3, 4 with the Surface of a Hypothetical Crystal Having a Characteristic Temperature of 510°K (Э = Translational Kinetic Energy of Molecules; T = Crystal Temperature)

m = 2			m = 3			m = 4		
Э, cal	T, °K 30	T, °K 100	Э, cal	T, °K 30	T, °K 100	Э, cal	T, °K 30	T, °K 100
1.49	0.1094	0.1310	0.99	0.0566	0.0707	0.74	0.0736	0.0966
5.95	0.0785	0.0939	3.95	0.0800	0.1008	2.98	0.0865	0.1139
23.8	0.0749	0.0893	15.8	0.0872	0.1090	11.9	0.0896	0.1174
53.55	0.0765	0.0907	35.5	0.0888	0.1104	26.78	0.0916	0.1198
95.2	0.0829	0.0971	63.2	0.0930	0.1150	47.6	0.0942	0.1225
214.2	0.0897	0.1009	142.2	0.0982	0.1175	107.1	0.0998	0.1259
380.8	0.0859	0.0911	252.8	0.0975	0.1103	190.4	0.1000	0.1201
595.0	0.0607	0.0621	395	0.0795	0.0803	297.5	0.0885	0.0992
856.8	0.0168	0.0170	568.8	0.0496	0.0511	428.4	0.0647	0.0690
1166.2	—		774.2	0.0236	0.0236	583.1	0.0384	0.0396
1523.2	—		1011.1	—		761.6	0.0152	0.0155
1927.8	—		1279.8	—		963.9	0.0028	0.0028

In this elementary treatment we have not allowed for the possibility that an evaporating particle can have some potential energy. The case to be considered is that in which a particle has departed from its equilibrium position by a distance $(x^2+y^2+z^2)^{1/2}$, so that to overcome the force of attraction it needs an energy smaller than the heat of evaporation by the amount

$$m/2 \cdot 4\pi^2\nu^2 (x^2+y^2+z^2).$$

The fraction of molecules whose departures from the position of equilibrium lie in the ranges x to x + dx; y to y + dy; z to z + dz, is given by

$$\left(2\pi\nu \sqrt{\frac{m}{2\pi k_B T}}\right)^3 e^{-\frac{m}{2}\frac{4\pi^2\nu^2(x^2+y^2+z^2)}{k_B T}} dx\, dy\, dz.$$

TABLE 2.2. Accommodation Coefficients of Air for Various Surfaces

Surface	a_K Minimum	a_K Maximum
Smooth black varnish on bronze	0.88	0.89
Polished bronze	0.91	0.94
Mechanically finished bronze	0.89	0.93
Etched bronze	0.93	0.95
Polished cast steel	0.87	0.93
Mechanically finished steel	0.87	0.88
Cast steel treated with aqua regia	0.89	0.96
Polished aluminum	0.87	0.95
Mechanically finished aluminum	0.95	0.97
Aluminum treated with aqua regia	0.89	0.97

According to Eq. (2.2), the fraction of molecules evaporated is

$$\left(2\pi\nu\sqrt{\frac{m}{2\pi k_B T}}\right)^3 e^{-\frac{m}{2}\frac{4\pi^2\nu^2(x^2+y^2+z^2)}{k_B T}} dx\,dy\,dz\,\frac{N}{V}\sqrt{\frac{k_B T}{2\pi m}}\,e^{-\frac{1}{k_B T}\left[\frac{Q_0}{N}-\frac{m}{2}4\pi^2\nu^2(x^2+y^2+z^2)\right]}$$

$$=\left(2\pi\nu\sqrt{\frac{m}{2\pi k_B T}}\right)^3 \frac{N}{V}\sqrt{\frac{k_B T}{2\pi m}}\,e^{-\frac{Q_0}{RT}} dx\,dy\,dz. \tag{2.3}$$

After integrating over the entire volume we get

$$a=\sqrt{\frac{k_B T}{2\pi m}}\left(\nu\cdot\sqrt{\frac{2\pi m}{k_B T}}\right)^3 e^{-\frac{Q_0}{RT}}. \tag{2.4}$$

Here ν is the vibrational frequency of an atom around its position of equilibrium.

According to the reciprocity principle, a single act of evaporation must involve the expenditure of the same amount of energy as will be released in a single act of condensation. Therefore the heat of condensation is equal in absolute value to the heat of evaporation. It is reasonable to regard condensation as a process in which incident atoms colliding with vibrating atoms in the surface lose a considerable fraction of their kinetic energy, until they are no longer able to pull themselves away again from the atoms of the surface. When particles strike surfaces heated to high temperatures, they can be reflected with a change (a decrease or an increase) of their energy.

These remarks bring out the connection between condensation and evaporation processes and the energy exchanges at a surface.

Lennard-Jones and Devonshire (1936) determined by quantum statistical calculation the probability of condensation for several hypothetical pairs of molecules and surfaces (Table 2.1).

If the temperature of the surface is T_{sur}, and a molecule originally at the temperature T_{inc} strikes it and comes away at the temperature T_{ref}, then by Eq. (2.5) we define the so-called Knudsen thermal accommodation coefficient

$$a_K=\frac{T_{ref}-T_{inc}}{T_{sur}-T_{inc}}. \tag{2.5}$$

If there is no exchange of energy, then $T_{ref}=T_{inc}$ and $a_K=0$. For $T_{ref}=T_{sur}$ (complete exchange of energy) $a_K=1$. Thus a_K varies from 0 to 1, depending on how completely the energy is exchanged. The accommodation coefficient varies over a wide range for different combinations of surfaces and molecular flows. In Table 2.2 we give values of a_K for air on various surfaces according to H. S. Tsien (1946).

In the work of J. de Boer (1962, pages 36-38), it is shown that for neon on a clean tungsten surface at room temperature $a_K=0.07$, and for the same surface covered with oxygen $a_K=0.6$.

For hydrogen on shiny platinum $a_K=0.323$, and on platinum black $a_K=0.583$.

It can be shown (de Boer, 1962) that if the gas and the surface are at different temperatures (a nonequilibrium condition) the actual pressure p' is related to the equilibrium pressure p by the formula

$$p' = \frac{1}{2} p \left[1 + \sqrt{1 + \frac{a_K (T_{sur} - T_{inc})}{T_{inc}}} \right]. \qquad (2.6)$$

If $T_{sur} > T_{inc}$, then $p' > p$; if $T_{sur} < T_{inc}$ then $p' < p$. From de Boer's formula we can derive a relation between the accommodation coefficient a_K and the evaporation coefficient α. If $T_{sur} < T_{inc}$, we can write

$$p' = \alpha p; \qquad (2.7)$$

and then

$$\alpha = \frac{1}{2} \left[1 + \sqrt{1 + \frac{a_K (T_{sur} - T_{inc})}{T_{inc}}} \right]. \qquad (2.8)$$

Owing to the fact that an actual surface has a more or less well developed texture (see section 1), a particle may have repeated collisions with different portions of the surface before it leaves the surface. Suppose that on the average the particle makes j such collisions; then the measured accommodation coefficient (Massey and Burhop, 1956) a_K will be related to the accommodation coefficient for an ideally smooth surface a'_K by the formula

$$a_K = 1 - (1 - a'_K)^j. \qquad (2.9)$$

To determine accommodation coefficients we can use (Roberts, 1939, 1940, 1950) a measurement of the energy loss $\mathcal{E}$ from a wire heated to a temperature T_2 inside a tube containing the gas to be studied. If the temperature of the walls of the tube is T_1, the accommodation coefficient is given by the formula

$$a_K = \frac{(2\pi m k_B T_1)^{1/2} \mathcal{E}}{2p(T_2 - T_1) k_B}, \qquad (2.10)$$

where p is the pressure of the gas in the tube.

If the molecules are polyatomic, then by introducing the respective accommodation coefficients a_K^t, a_K^r, a_K^v for translational, rotational, and vibrational motions, respectively, we can write for the energy loss

TABLE 2.3. Equilibrium Temperature T_s of the Surface of a Spherical Artificial Satellite for Various Speeds and Heights of Its Motion During the Daytime Part of the Orbit

Height, km	n, g/cm³	$\frac{a_K}{\varepsilon}$	T_s, °K		
			2000 m/sec	4000 m/sec	8000 m/sec
72	10^{-7}	0.15	720	1240	2023
86	10^{-8}	0.5	554	917	1537
98	10^{-9}	1	395	619	1029
116	10^{-10}	1.5	313	409	646
134	10^{-11}	2	293	314	414

$$\vartheta = \frac{p(4a_K^t + f_r a_K^r + f_v a_K^v)(T_2 - T_1)\,k_B}{2(2\pi m k_B T)^{1/2}}; \qquad (2.11)$$

f_r and f_v are the fractions of the energy which belong to the rotational and vibrational motions.

The theory of the accommodation coefficient of an atom on a metal surface has been developed by Jackson and Mott, 1932 (quoted by Massey and Burhop, 1956). The formula derived by L. D. Landau in 1935 for a monatomic gas is

$$a_K = \frac{3}{8\,m_1\,m^{1/2}}\left(\frac{2h^2T}{\pi^2 k_B d^2 \theta^2}\right)^{3/2}, \qquad (2.12)$$

where m and m_1 are the masses of the gas atom and an atom of the solid, θ is the characteristic temperature of the solid, and, as in Eq. (2.4), d is the width of the region of interaction and is of the order of 10^{-8} cm. In 1937, Devonshire found the limits of validity of Eq. (2.12) and made theoretical calculations which hold over a wider temperature range and for a larger number of pairs of substances than Eq. (2.12).

The state and behavior of a surface are described by a large number of coefficients of all kinds: coefficients of reflection, slip, momentum transfer, accommodation of rotational, translational, and vibrational energy, condensation, evaporation, and recombination; there are also various coefficients of chemical activity, emittance, etc.

Let us consider the coefficient of momentum transfer. Let the component of the velocity of a particle parallel to the plane of a plate be u_x'' (after reflection this same component is u_x^o). Then the momentum transfer coefficient (Maxwell coefficient) is defined by the equation

$$f_M = \frac{mu_x'' - mu_x^o}{mu_x''} = \frac{u_x'' - u_x^o}{u_x''}. \qquad (2.13)$$

In the case of specular reflection $u_x'' = u_x^o$ and $f_M = 0$. If the surface is sufficiently rough, then $u_x^o = 0$, and $f_M = 1$, this last being true only on the average.

We have already given an example of a connection between the different coefficients describing a surface [Eqs. (2.8), (2.9)]. Relations between the coefficients are of great importance in practical applications.

To conclude this section we give an example of the connection between the emittance and the accommodation coefficient. The basic results are due to M. Devienne (1962).

TABLE 2.4. Equilibrium Temperature of the Surface of a Spherical Artificial Satellite for Various Speeds and Heights of Its Motion in Nighttime

Height, km	n, g/cm³	a_K*/ε	T_s°K 2000 m/sec	4000 m/sec	8000 m/sec
72	10^{-7}	0.15	715	1203	2023
86	10^{-8}	0.5	545	914	1537
98	10^{-9}	1	367	612	1028
116	10^{-10}	1.5	241	384	641
134	10^{-11}	2	183	244	389

*a_K/ε is assumed for nickel-plated copper sheet.

Let us consider the thermal balance at the surface of a spherical artificial satellite of radius R_S. Let a_K be the accommodation coefficient of the surface, independent of the angle of incidence of particles on the surface. We neglect the thermal velocities of the particles, the changes of the rotational and vibrational energies of the particles on impact, and also possible chemical reactions at the surface.

The energy received by the satellite in its collisions with gas particles is given by

$$Э = a_K n\pi R_S^2 u \cdot 0.5 m u^2 = 0.5\pi R_S^2 n u^3,$$

where u is the speed of the satellite. There are different energy balance equations for day and night.

I. For day:

$$4\pi R_S^2 \varepsilon \sigma T_S^4 = 0.5 a_K \pi R_S^2 n u^3 + \pi R_S^2 \varepsilon (E_s + E_e),$$

where E_s and E_e are the solar and earth irradiation constants of the satellite. $E_s = 14.10^5$ erg cm^{-2} sec^{-1}, $E_e = 2.10^5$ erg cm^{-2} sec^{-1} by day and $E_e = 1.75 \cdot 10^5$ erg cm^{-2} sec^{-1} by night; ε is the emittance.

II. For night:

$$4\pi R_S^2 \varepsilon \sigma T_S^4 = 0.5 a_K \pi R_S^2 n u^3 + \pi R_S^2 \varepsilon E_e.$$

Because the theory of the relation between the various coefficients is not very well developed as yet, we confine ourselves to the somewhat fragmentary information which is available.

SECTION 3

THE MOTION OF RAREFIED GASES THROUGH CYLINDRICAL TUBES (I)

(Fundamental Results of the Theory of the Transfer of Momentum from a Gas to the Walls of a Tubular Channel)

In its initial stage the theoretical treatment of the flow of highly rarified gases through tubes was conducted by taking into account the transfer of momentum from the gas to the wall. We shall list the main results obtained by Knudsen, Smoluchowski, and Dushman (given in Clausing, 1932), and shall present in more detail the calculations which have been made relatively recently by Pollard and Present (1948).

In M. Knudsen, 1909 (1) the following formula is given for the stationary flow of a highly rarefied gas through a cylindrical tube with cross section of arbitrary shape:

$$J_{pV} = \frac{8}{3}\sqrt{\frac{2}{\pi}}\,\frac{S^2(p_1 - p_2)}{sL\sqrt{d}}, \tag{3.1}$$

where J_{pV} is the quantity of gas flowing through per second, d is the density of the gas at unit pressure, S is the area of the cross section, s is the perimeter of the cross section, and p_1 and p_2 are the pressures at the two ends of the cylinder in dynes/cm^2.

The formula (3.1) is valid if, (1), the length L is much larger than the transverse dimension of the tube, (2), the mean free path of the particles is of the same order of magnitude as the transverse dimension of the tube, and (3), the molecules are scattered at the walls of the tube according to the cosine law.

In particular, for a long circular cylindrical tube of radius r we can write Eq. (3.1) in the form

$$J_{pV} = \frac{4\sqrt{2\pi}}{3}\,\frac{r^3}{L\sqrt{d}}(p_1 - p_2). \tag{3.2}$$

For the other extreme case of flow through a very short cylinder (an aperture in a thin wall) we find in M. Knudsen, 1909 (2) the suggested formula

$$J_{pV} = \frac{S}{\sqrt{2\pi}}\,\frac{1}{\sqrt{d}}(p_1 - p_2), \tag{3.3}$$

from which we get for the special case of a circular aperture

$$J_{pV} = \sqrt{\frac{\pi}{2}}\,\frac{r^2}{\sqrt{d}}(p_1 - p_2). \tag{3.4}$$

Smoluchowski (1910) shows, however, that it is more correct to write Eq. (3.1) in the form

$$J_{pV} = \frac{1}{2\sqrt{\pi}}\,\frac{A}{L\sqrt{d}}(p_1 - p_2), \tag{3.5}$$

where $A = \frac{1}{2}\int\limits_{s}\int\limits_{-\frac{\pi}{2}}^{\pi/2} k^2 \cos\theta d\theta ds$ and k is the cross-sectional chord forming an angle θ with the normal to ds. It is easy to show that Eqs. (3.2) and (3.5) are valid only for comparatively long tubes, since when we go to z = 0 both formulas would give $J_{pV} = \infty$, instead of going over into Eq. (3.3).

Dushman (1922, 1926, 1931) gave a formula by means of which one can obtain this limiting transition to short tubes.

It was shown in Clausing (1932) that these results of Dushman must be regarded as only crude approximations. In fact, Dushman's arguments are open to objections. He rewrites the expressions (3.2) and (3.4) in the form:

$$J_{pV} = \frac{p_1 - p_2}{\omega'}; \quad \omega' = \frac{3}{4\sqrt{2\pi}} \frac{L}{r^3} \sqrt{d}, \tag{3.6}$$

$$J_{pV} = \frac{p_1 - p_2}{\omega''}; \quad \omega'' = \sqrt{\frac{2}{\pi}} \frac{1}{r^2} \sqrt{d}; \tag{3.7}$$

here ω' and ω" are the amounts of resistance presented to a molecular flow by a tube and by an aperture.

If the resistances are in series, we can write for the total resistance

$$\omega = \omega' + \omega''. \tag{3.8}$$

We thus have

$$J_{pV} = \frac{p_1 - p_2}{\omega' + \omega''}. \tag{3.9}$$

The application of Eq. (3.3) to short tubes is not permissible, since it is valid only for long tubes.

Let u_x be the average velocity of the molecules along the axis of the tube (in other words, this is the average velocity of mass motion of the gas along the axis of the tube).

The frequency of collisions of gas molecules with an elementary band of width dx around a tube of radius r is given by

$$\frac{1}{4} nu \cdot 2\pi r dx = \frac{1}{2} nu\pi r dx.$$

The rate of momentum transfer is $\frac{1}{2} n\pi rumu_x dx$. We equate this rate of momentum transfer to a force $\pi r^2 dp$ produced by the difference of pressure between the sections x and x + dx; neglecting the force of inertia, which is not very large for long tubes, we get

$$nmu_x = -\frac{2r}{u}\frac{dp}{dx}.$$

The mass of gas J_m which passes through the tube per unit time is given by

$$J_m = Smnu_x = -\frac{2Sr}{u} \cdot \frac{dp}{dx}; \tag{3.10}$$

or the number of particles per unit time is

$$J = \frac{J_m}{m} = -\frac{2Sr}{um}\frac{dp}{dx}.$$

For a **circular tube**, $S = \pi r^2$ and

$$J = -\frac{2\pi r^3}{um} \cdot \frac{dp}{dx}. \tag{3.11}$$

An analogous calculation using Smoluchowski's Eq. (3.5) gives for a circular tube a value which better fits the experimental data:

$$J = -\frac{16}{3}\frac{r^3}{mu}\frac{dp}{dx}. \tag{3.12}$$

For cases of great practical importance in which the mean free path $\lambda \ll r$, it is important to distinguish between two different forms for J: (1) hydrodynamic viscous flow, and (2) gaseous self-diffusion flow.

The first part of the flow is determined mainly by the gradient of the total pressure along the tube, and is given by the usual hydrodynamic expression [cf., Kennard (1938, page 293)]:

$$J = n\pi r^2 \left[-\frac{r^2 dp}{8\eta\, dx} + u_0\right]. \tag{3.13}$$

The theory of hydrodynamic flow, not taking into account molecular drift velocity, is usually confined to calculations of the viscosity η and the rate of slip u_0 along the walls. According to Maxwell,

$$u_0 = -\frac{2\eta}{nmu} \cdot \left(\frac{du}{dr}\right)_{r=a} \tag{3.14}$$

$$\eta = \frac{1}{3} nmu\lambda, \tag{3.15}$$

$$p = \frac{\pi}{8} nmu^2. \tag{3.16}$$

In the expression (3.15) we have used a factor $^1/_3$ instead of $^1/_2$, in accordance with the data of Chapman and Enskog (quoted by Pollard and Present, 1948). Using Eqs. (3.14)-(3.16), we can write Eq. (3.13) in the form

$$J = -\left[\frac{2\pi r^3}{mu} + \frac{3\pi r^4}{8mu\lambda}\right]\frac{dp}{dx}. \tag{3.17}$$

In experiments by Knudsen (1909) (1), Gaede (1913), and Adzumi (1937) the transition from the range of pressures in which Eq. (3.12) is valid to the region where Eq. (3.13) is valid was investigated experimentally. At low pressures the expression (3.14) goes over into the same form as the formula for free molecular flow (3.12), but is smaller by a factor $3\pi/16$. This factor does not depend on the quantities η and λ. If we integrate Eq. (3.17) and plot the specific flux $J/\Delta p$ against the average pressure, we get a straight line with an intercept on the ordinate axis corresponding to the slip term in Eq. (3.17).

Since the intercept lies below the line for free molecular flow, the curve of the specific flux must have a minimum near the point $r/\lambda = 0.2$. A minimum is actually observed experimentally, but it lies at smaller values of r/λ and is not very pronounced.

The classical treatment of slip flow cannot, however, guarantee the exact numerical value of the factor $3\pi/16$, and the prediction of a minimum in the fundamental equation (3.17) is essentially rather arbitrary.

In the extrapolation to zero pressure, the experimental curves correspond within one or two percent to the formula (3.12) of Knudsen and Smoluchowski. Gaede's experiments showed that Eq. (3.12) is obeyed for $\lambda/r \approx 100$. An explanation of these effects will be given below.

The problem which arises in the other limiting case of high pressures with a small total pressure gradient is that of the self-diffusion of the gas particles. This problem can be approached experimentally if one labels some of the gas molecules. The transfer caused by internal diffusion is connected with the partial pressure gradient of labeled molecules by the relation

$$J_1 = -\frac{\pi r^2}{k_B T} D_{11} \frac{dp}{dx}, \tag{3.18}$$

where p is the partial pressure of labeled molecules and D_{11} is their coefficient of self-diffusion.

When there is no total pressure gradient and the molecular drift by diffusion transfer is extremely slow, it is possible to consider theoretically the transition from Eq. (3.18), which is valid at extremely large pressures. Let $a = J/\pi r^2$, and define the total diffusion coefficient by the relation

$$a = -D \frac{dn}{dx}. \tag{3.19}$$

If we remember that

$$p = nk_B T = \frac{\pi n m u^2}{8}, \tag{3.20}$$

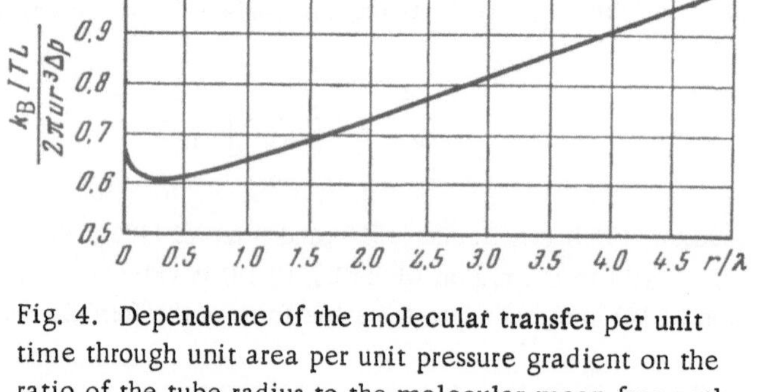

Fig. 4. Dependence of the molecular transfer per unit time through unit area per unit pressure gradient on the ratio of the tube radius to the molecular mean free path (Pollard and Present).

then in Eq. (3.19) we have

$$D=\frac{2}{3}ur, \tag{3.21}$$

which holds for $r/\lambda \to 0$. Elementary kinetic theory gives the following expression for the coefficient of self-diffusion:

$$D=\frac{u\lambda}{3}, \tag{3.22}$$

which holds for $r/\lambda \to \infty$.

At low pressures, when $L>\lambda>>r$, an average molecule crossing the tube can only have a collision with another molecule by accident. The small component of flow which can appear under these conditions is proportional to the probability of collisions between molecules – that is, to r/λ.

At zero pressure ($\lambda>>L$) the transport consists of Knudsen diffusion. When the pressure is raised somewhat ($L>\lambda>r$), the flow tends to decrease a bit because of the decreased mean free path, but on the other hand tends to increase because of the increased transport by flow. Therefore the Knudsen flux must first fall with the pressure, go through a minimum, and then increase as the pressure is raised further, approaching the Poiseuille law. This sort of behavior is observed experimentally, as can be seen from the curve in Fig. 4. This curve was obtained by plotting the Knudsen specific flux for a number of different gases and capillaries, using the quantity $k_B T l L/2\pi u r^3 \Delta p$ as ordinate and r/λ as abscissa. Here Δp is the pressure drop along the capillary.

To get a more accurate idea of the limits of applicability of the concepts developed here, we give an approximate formula for the dependence of the mean free path λ on the pressure p:

$$\lambda \cong \frac{10^{-5}}{p}, \tag{3.23}$$

where λ is in cm and p is in atm. It follows from Eq. (3.23) that in capillaries with diameters of the order of 100 A the Knudsen flow persists up to pressures of about 10 atm.

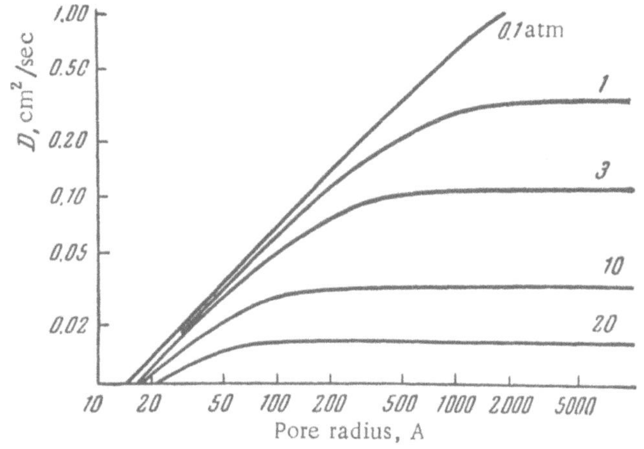

Fig. 5. Dependence of diffusion coefficient on the size of the capillary and on the pressure (Wheeler).

E. Wheeler (1951) has proposed two semiempirical formulas which in the limiting cases of small and large pressures give, respectively, for the diffusion coefficient

$$D = \frac{1}{3} u\lambda (1 - e^{-2r/\lambda}) \tag{3.24}$$

and

$$D = \frac{1}{3} u\lambda \frac{2r}{\lambda + 2r}. \tag{3.25}$$

Figure 5 shows the dependence of the diffusion coefficient on the radius of the capillary and on the pressure. The calculations were made with Eq. (3.24) using the value $u = 10^5$ cm/sec and the value of λ given by Eq. (3.23). Accordingly Fig. 5 holds for "average" molecules and average temperatures. When one needs to know D for mixtures of gases, the data of Fig. 5 must be recomputed by means of Eq. (3.24) using the actual values of u and λ (this is not always a simple task).

So far we have been considering the motion of a gas along a strictly isothermal capillary. Knudsen (1910) (7) made a theoretical and experimental study of the problem of the pressure difference which appears at the end of a nonuniformly heated capillary. For this case he found the differential equation

$$\frac{dp}{dT} = \frac{p}{T} f\left(\frac{r}{\lambda}\right), \tag{3.26}$$

where f is a function which depends on the ratio of the mean free path to the radius of the tube.

SECTION 4

MOLECULAR FLOW INSIDE A CYLINDER

In a paper by Sumpner (1892, quoted by Walsh, 1919-1920) it is shown that if the inside surface of a sphere (see Fig. 6) consists of diffusive areas (i.e., areas obeying the cosine law), then the flux from any surface element dS is distributed in the same way among the other surface elements dS', independent of the relative positions of dS and dS' on the surface of the sphere.

Let us consider the problem of the amount of radiation (heat or matter) which passes from a diffusely radiating (evaporating) disk to a parallel and coaxial disk (Walsh, 1919-1920).

In the sphere of Fig. 6 let there be placed a flat disk AB which radiates according to the cosine law. The radiation at any point of this disk will be the same as the radiation from the segment of the sphere ACB on which the disk rests.

Accordingly, the fraction of the total flux from AB which falls on A'B' (the parallel disk) is equal to the ratio of the spherical area A'B'C' to the entire area AA'C'B'B. If R is the radius of the sphere and L the distance between AB and A'B', and if A'B' = 2r, AB = 2r', then

$$L = \sqrt{R^2 - r^2} + \sqrt{R^2 - r'^2}.$$

The calculation of the spherical areas S(A'C'B') and S(AA'C'B'B) gives the results

$$S(A'C'B') = 2\pi R\,(R - \sqrt{R^2 - r^2}),$$

$$S(AA'C'B'B) = 2\pi R\,(R + \sqrt{R^2 - r'^2}).$$

If E_0 is the flux emitted normal to the surface of the disk AB per unit time and unit area, then the total flux radiated by the disk AB will be $\pi r'^2 \pi E_0$, and the number of particles E (cm^{-2} sec^{-1}) impinging on A'B' is

$$E = \frac{R - \sqrt{R^2 - r^2}}{R + \sqrt{R^2 - r'^2}}\, \pi^2 r'^2 E_0$$

and we have finally

$$E = \frac{\pi^2 E_0}{2}\,[L^2 + r^2 + r'^2 - \sqrt{(L^2 + r^2 + r'^2)^2 - 4r^2 r'^2}]. \quad (4.1)$$

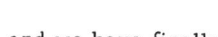

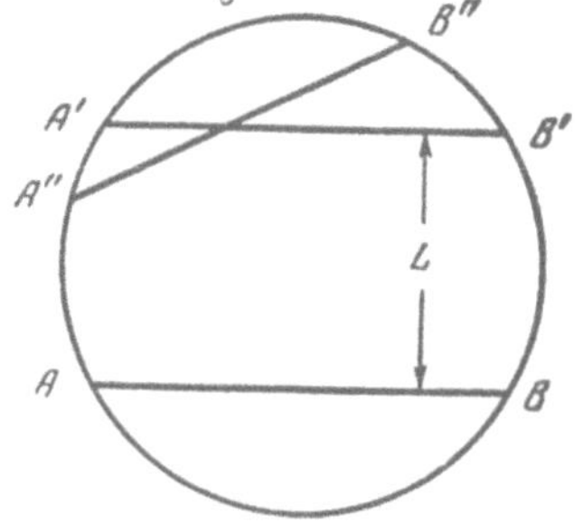

Fig. 6. Spherical cavity at equilibrium.

From what has been said it can be concluded that the same amount of radiation will also be received by any other disk A"B" of the same radius r as A'B' and placed so that its edge is the circumference of a small circle on the surface of the sphere.

This same problem was solved by very cumbersome quadruple integrals by Owen (1920), and then in a paper by Bartlett (1920) by exactly the same method as used by Walsh. In 1924 V. A. Fok published a paper, "The Luminance from a Surface of Arbitrary Shape," wherein he derived a number

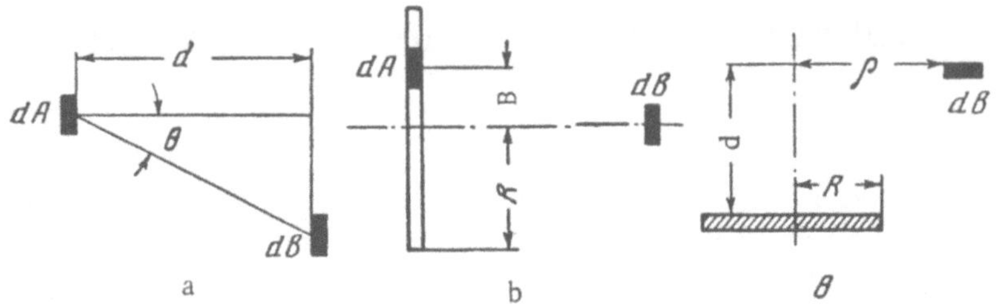

Fig. 7. Derivation of the formulas of Walsh, Hyde, and Foote.

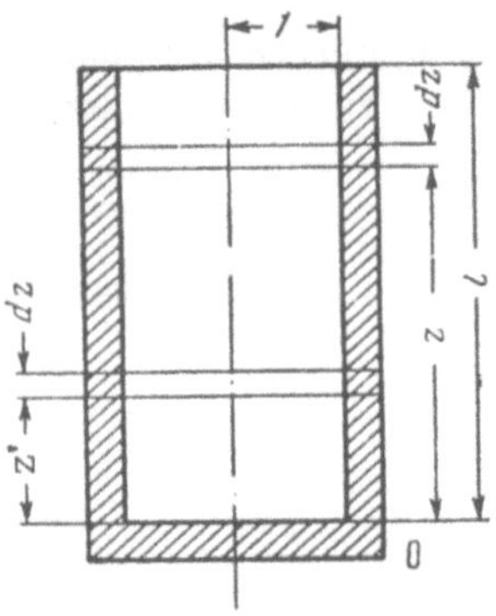

Fig. 8. Calculation of the distribution of molecular flows in horizontal layers in a cylinder.

of general relations, from which the expression (4.1) can be obtained as a special case.

We present without proof a few more results of Walsh, which, as he himself said, had been obtained earlier by other authors.

Let dA be an element of a perfectly diffusive surface (see Fig. 7a). If dB is an element of a surface parallel to dA and located at the normal distance d from dA, then the intensity of the radiation which falls on dB from dA is

$$E = \frac{E_0 \cos^4 \theta \, dA \, dB}{d^2}. \tag{4.2}$$

From Eq. (4.2) we find the flux which is received from a radiating disk by an elementary area which is parallel and coaxial with it (see Fig. 7b):

$$E = \frac{E_0 \pi R^2 \, dB}{R^2 + d^2}. \tag{4.3}$$

The result (4.3) had been obtained earlier by Hyde (1907).

We give the formula for the flux (see Fig. 7c) received from a disk by an elementary area dB which is parallel to the disk but located at a distance ρ' from the axis (d is the normal distance and R is the radius of the disk):

$$E = \frac{E_0 \pi}{2} \left[1 - \frac{c}{\sqrt{c^2 + R^2}} \right] dB, \tag{4.4}$$

where

$$c = (d^2 - R^2 + \rho'^2)/2d.$$

Equation (4.4) was found by Foote in 1916.

The results we have obtained are quite sufficient for calculation of the distribution of molecular flows inside a diffusely evaporating and scattering cylinder.

Let us consider the cylinder shown in Fig. 8. Let the coefficients of evaporation from the bottom and from the walls be α and β respectively. Then for disks of the cylinder which are located at distances z and z' from the bottom we can write the Walsh formula (4.1) in the form

$$E = \frac{\pi a}{2} F(z), \tag{4.5}$$

where $a = E_0\pi$ is the amount of matter which evaporates per unit time from unit area of the base of the disk, and F(z) is given by the formula:

$$F(z) = z^2 + 2 - z\sqrt{z^2 + 4}. \tag{4.6}$$

Accordingly, $0.5\,\pi a F(z)$ molecules ($cm^{-2}\,sec^{-1}$) will strike a disk at the height z and $0.5\pi a\,F(z + dz)$ molecules ($cm^{-2}\,sec^{-1}$) will strike a disk at the height z+ dz after having emerged from the base.

Consequently,

$$\frac{\pi a}{2}[F(z) - F(z + dz)] = -\frac{\pi a}{2}\frac{dF(z)}{dz}dz$$

molecules strike within the band between z and z+dz. Since the area of this band is $2\pi dz$, then there are $-(a/4)\,dF(z)/dz$ molecules which strike the band at the height z per unit time and unit area. From this we see that if the rate of evaporation of the walls of the cylinder is b, then the total number of molecules that leave a unit area of this band per unit time at the height z is $\varphi_0(z)$, given by:

$$\varphi_0(z) = b - (1 - \beta)\frac{a}{4}\frac{dF}{dz}. \tag{4.7}$$

If we assume that φ_0 remains constant between z and z+ dz, then the number of molecules received by the disk at z from the band at z will be:

$$\frac{\pi\varphi_0}{2}\{F|z - z_1| - F|(z + dz) - z_1|\},$$

or

$$-\frac{\pi\varphi_0}{2}\frac{dF|z - z_1|}{dz}dz.$$

Accordingly the number of molecules received per unit area at z_1 from the band at z is

$$0.25\,\varphi_0\frac{d^2F|z - z_1|}{dz^2}dz.$$

The number of particles which leave z_1 after being distributed over the horizontal bands φ_0 of the cylinder will be:

$$\varphi_1 = 0.25(1 - \beta)\int_0^{z_1}\varphi_0\frac{d^2}{dz^2}dz + 0.25(1 - \beta)\int_{z_1}^{l}\varphi_0\frac{d^2F(z - z_1)}{dz^2}dz$$

$$+ 0.125(1 - \alpha)(1 - \beta)\frac{dF(z_1)}{dz}\int_0^{l}\varphi_0\frac{dF}{dz}dz. \tag{4.7'}$$

Similar expressions can be written for the quantities $\varphi_2, \varphi_3 \ldots \varphi$, and therefore the number of particles which leave z after n-fold reflection from the walls of the cylinder is

$$\Phi(z_1) = \sum_{i=0}^{n} \varphi_i = \varphi_0 + 0.25(1-\beta)\int_0^{z_1} \Phi(z)\frac{d^2F(z_1-z)}{dz^2}dz$$

$$+0.25(1-\beta)\int_{z_1}^{l} \Phi(z)\frac{d^2F(z-z_1)}{dz^2}dz$$

$$+0.125(1-\alpha)(1-\beta)\frac{dF(z)}{dz}\int_{z_1}^{l} \Phi(z)\frac{dF}{dz}dz. \quad (4.8)$$

The integral equation (4.8) is the fundamental equation for our further exposition.

Simplified versions of this equation in photometry were derived by Buckley (1927, 1928, 1934) and in the theory of molecular flows by Clausing [1929, (4)].

SECTION 5

THE MOTION OF RAREFIED GASES THROUGH CYLINDRICAL TUBES (II)

Here we introduce the following definitions (cf. Clausing, 1932):

1. $\omega_{bb}(z)$ is the probability that a molecule emitted from the band $2\pi\, rdz$ in accordance with the cosine law will reach another band located at a distance z from the first band without any earlier collision with the wall.

2. $\omega_{bd}(z)$ is the probability that a molecule emitted according to the cosine law from the band $2\pi\, rdz$ will pass through a disk located at the distance z from the band.

3. $\omega_{db}(z)$ is the average over the disk πr^2 of the probability that a molecule emitted according to the cosine law from the disk will fall on a band $2\pi\, rdz$ located at a distance z from the disk.

4. $\omega_{dd}(z)$ is the average over the disk πr^2 of the probability that a molecule coming from the disk will arrive at another disk located at a distance z from the first one.

It is easily verified that

$$w_{bb} = -\frac{dw_{bd}}{dz}, \tag{5.1}$$

$$w_{db} = -\frac{dw_{dd}}{dz}, \tag{5.2}$$

$$w_{db} = 0{,}5rw_{bd} \tag{5.3}$$

Knowing the formulas (5.1)-(5.3) and, for example, ω_{dd}, it is easy to find the other probabilities; ω_{dd} is determined from the formula of Walsh derived above:

$$w_{dd} = \frac{1}{2r^2}\{z^2 - z\sqrt{z^2+4r^2} + 2r^2\}. \tag{5.4}$$

From Eqs. (5.4) and (5.1)–(5.3) we have the formulas

$$w_{db}(z) = \frac{1}{2r^2}\left[\sqrt{z^2+4r^2} + \frac{z^2}{\sqrt{z^2+4r^2}} - 2z\right], \tag{5.5}$$

$$w_{bd}(z) = \frac{1}{4r}\left[\sqrt{z^2+4r^2} + \frac{z^2}{\sqrt{z^2+4r^2}} - 2z\right], \tag{5.6}$$

$$w_{bb}(z) = \frac{1}{4r}\left[2 + \frac{z^3}{(\sqrt{z^2+4r^2})^3} - \frac{3z}{\sqrt{z^2+4r^2}}\right]. \tag{5.7}$$

For $z \gg 2r$ we have the convergent series

$$w_{dd} = \frac{1}{4}\left(\frac{2r}{z}\right)^2 - \frac{1}{8}\left(\frac{2r}{z}\right)^4 + \frac{5}{64}\left(\frac{2r}{z}\right)^6 - \ldots. \tag{5.8}$$

TABLE 5.1. Probability of Transmission of a Molecule Through a Tube as a Function of the Ratio L/r (Clausing Coefficient)

L/r	W	L/r	W	L/r	W
0	1	1.7	0.5518	8	0.2316
0.1	0.9524	1.8	0.5384	9	0.2131
0.2	0.9092	1.9	0.5226	10	0.1973
0.3	0.8699	2.0	0.5136	12	0.1719
0.4	0.8341	2.2	0.4914	14	0.1523
0.5	0.8013	2.4	0.4711	16	0.1367
0.6	0.7711	2.6	0.4527	18	0.1240
0.7	0.7434	2.8	0.4359	20	0.1135
0.8	0.7177	3.0	0.4205	30	0.0797
0.9	0.6940	3.2	0.4062	40	0.0613
1.0	0.6720	3.4	0.3931	50	0.0499
1.1	0.6514	3.6	0.3809	60	0.0420
1.2	0.6320	3.8	0.3695	70	0.0363
1.3	0.6139	4.0	0.3589	80	0.0319
1.4	0.5970	5	0.3146	90	0.0285
1.5	0.5810	6	0.2807	100	0.0258
1.6	0.5659	7	0.2537	1000	0.002658

The probability W that a molecule passing through a vessel of length L will enter a second vessel and not return to the first one will be called the probability of transmission. W is defined by the equation

$$W = w_{dd}(L) + \int_0^L w_{db}(z)\, W(z)\, dz. \tag{5.9}$$

The integral equation (5.9) is very similar in form to the integral equation derived at the end of Section 4. The difference between them is due to the fact, first, that Clausing does not take into account the evaporation coefficient, and second, that there is no bottom to Clausing's cylinder. Other relations are also easily derived:

$$W(z_1) = w_{bd}(L - z_1) + \int_0^L w_{bb}(z - z_1)\, W(z)\, dz \tag{5.10}$$

or, using Eqs. (5.6) and (5.7),

$$W(z) = \frac{1}{4r}\left[\sqrt{(L-z)^2 + 4r^2} + \frac{(L-z)^2}{\sqrt{(L-z)^2+4r^2}} - 2(L-z)\right]$$

$$+ \frac{1}{4r}\int_0^L \left[2 + \frac{(z_1 - z)^3}{\sqrt{(z_1-z)^2+4r^2}} - \frac{3(z_1-z)}{\sqrt{(z_1-z)^2+4r^2}}\right] W(z_1)\, dz_1. \tag{5.11}$$

Substituting, we get

$$W(z) = \gamma + \frac{1-2\gamma}{L} z, \tag{5.12}$$

where

$$\gamma = \frac{\Phi(0)}{a}.$$

After extremely tedious and cumbersome calculations, Clausing derived formulas for short and long tubes. If we denote the cross-sectional area of the tube by S, then

$$J = WSa. \tag{5.13}$$

Clausing gives a table of values of $W = W(L/r)$. By means of Eq. (5.13) and the data of Table 5.1 it is now easy to calculate J.

Clausing also found values of W for narrow slotted tubes. If b is the width of the tube, H is its height, and L its length, then under the conditions $b \ll H$ and $L < H$ the values for W are given by the following table (Table 5.2). The formulas are omitted in view of their considerable bulkiness.

To conclude this section we consider examples of the use of the relations we have given for calculations in high-vacuum technology (Clausing, 1932, translated into Russian by G. D. Kuznetsova):

There is a well known rule: One must make the cross section of the tubes as large as possible and the length as small as possible. The resistance to flow is proportional to the length and inversely proportional to the third power of the cross-sectional diameter.

There is very little to add to this in principle. The author would like to illustrate this rule with a table which gives the time for pumping out tanks using a given type of tube linkage.

Let us consider a tank of volume V which is connected with tubes to a high-vacuum pump. At the time $t = 0$ the gas in the reservoir is at a pressure $p_1 = p_{10}$ and is being removed by the pump. The instantaneous pres-p_1 satisfies the equation

$$-V dp_1 = J\, dt \tag{5.14}$$

TABLE 5.2. Clausing Coefficient for Slotted Tubes

L/b	W	L/b	W	L/b	W
0	1	1.3	0.6321	3.2	0.4439
0.1	0.9525	1.4	0.6168	3.4	0.4318
0.2	0.9096	1.5	0.6024	3.6	0.4205
0.3	0.8710	1.6	0.5888	3.8	0.4099
0.4	0.8362	1.7	0.5760	4.0	0.3999
0.5	0.8048	1.8	0.5640	5.0	0.3582
0.6	0.7763	1.9	0.5525	6.0	0.3260
0.7	0.7503	2.0	0.5417	7.0	0.3001
0.8	0.7266	2.2	0.5215	8.0	0.2789
0.9	0.7049	2.4	0.5032	9.0	0.2610
1.0	0.6848	2.6	0.4865	10.0	0.2475
1.1	0.6660	2.8	0.4712	∞	$(b/L)\ln(L/b)$
1.2	0.6485	3.0	0.4570		

or, using Eq. (3.9),

$$-V\,dp_1 = \frac{p_1 - p_2}{\omega}\,dt, \tag{5.15}$$

where p_2 is the pressure in the pump. On the other hand, p_2 satisfies the equation

$$-V\,dp_1 = D_p p_2\,dt, \tag{5.16}$$

where D_p is the pumping speed.

Eliminating p_2 from Eqs. (5.15) and (5.16), we get:

$$-V\,dp_1 = \frac{p_1}{\frac{1}{D_p} + \omega}\,dt. \tag{5.17}$$

Integration of Eq. (5.17) gives

$$t = V\left(\frac{1}{D_p} + \omega\right)\ln\frac{p_{10}}{p_1}. \tag{5.18}$$

The conduit system linking the tank to pump consists of a number of tubes of different diameters connected in series. A stopcock included in the system may be regarded as a short tube. To these various tubes there correspond different resistances ω', ω'', ω''', etc. Therefore we can put Eq. (5.18) in the form

$$t = V\left(\frac{1}{D_p} + \omega' + \omega'' + \ldots\right)\ln\frac{p_{10}}{p_1}. \tag{5.19}$$

It can be seen from Eqs. (5.18) and (5.19) that the reciprocal of the pumping speed may be regarded as a resistance.

We now imagine the following series of experiments:

1. The tank is connected directly to the pump without any transition tube; as before, the evacuation is from p_{10} to p_1 according to Eq. (5.18). The time necessary for this is

$$t_D = \frac{V}{D_p}\ln\frac{p_{10}}{p_1}. \tag{5.20}$$

2. The tank from which the air is pumped is connected by the first tube to an infinitely large evacuated volume. We again solve for the time necessary for the pressure to fall from p_{10} to p_1:

$$t' = V\omega'\ln\frac{p_{10}}{p_1}. \tag{5.21}$$

3. The next experiments are the same as the second experiment but with the use of a second tube, a third tube, and so on. For the second tube we get

$$t'' = V\omega''\ln\frac{p_{10}}{p_1}. \tag{5.22}$$

Using Eqs. (5.19)-(5.22), we can accordingly write:

$$t = t_D + t' + t'' + \ldots \tag{5.23}$$

Now let us proceed in a somewhat arbitrary way to use the evacuation time and the partial evacuation times to find those values of t, t_D, $t' \ldots t^{(i)}$ which correspond to the conditions $p_{10}/p_1 = 1000$ or $\ln p_{10}/p_1 = 6.9078$. When we also use Eq. (5.20), we can write

$$t_D = 6.9078V/D_p = V \cdot \tau_D, \tag{5.24}$$

while for the stopcocks and tubes system, according to Eq. (5.22)

$$t^{(i)} = V\omega^{(i)} 6.9078 = V\tau^{(i)}. \tag{5.25}$$

We define the quantities $\tau^{(i)}$ as the partial pumping times required for evacuating a unit volume. For the system of tubes, which can almost always be regarded as consisting of long tubes with resistance proportional to the length, we define the partial pumping time σ per unit volume and per unit length by means of the equation:

$$t^{(m)} = VL^{(m)} \frac{\omega^{(m)}}{L^{(m)}} 6.9078 = VL^{(m)}\sigma^{(m)}. \tag{5.26}$$

It is now clear that we can use the formula

$$t = V[\tau_D + \Sigma\tau^{(n)} + \Sigma L^{(m)}\sigma^{(m)}] \tag{5.27}$$

to determine the pumping time for any system of tubes if we possess a table of the values of τ and σ for the stopcocks and tubes which are used.

Tables 5.3 and 5.4 give data of this sort for nitrogen being pumped at 18°C. In using the tables, one must substitute into Eq. (5.27) the value of $L^{(m)}$ in meters and V in liters to get the pumping time in seconds. The quantity σ in the tables is calculated by means of Eq. (5.29), which follows from the formulas we have given: Eq. (5.26), the second of Eqs. (3.6), and the formula for the density at unit pressure:

$$d = \frac{8}{\pi u^2}, \tag{5.28}$$

TABLE 5.3. (ϕ is the Orifice Diameter of the Stopcock in mm, L is the Length of the Stopcock Orifice in mm, and τ is the Pumping Time for Nitrogen in sec/liter)

Description of stopcock	ϕ	L	τ
Three-way stopcock with one orifice	6	1.5	2.6
Two-way stopcock with two orifices	6	2 × 1.5	5.2
Two-way stopcock with two orifices	10	2 × 2	1.8
Three-way stopcock with one orifice	5	1.5	3.98
Two-way stopcock	2	9	88.0
Two-way stopcock	6	15	6.62

TABLE 5.4. (ϕ is the Orifice Diameter in mm, σ is the Partial Pumping Time for Nitrogen in sec/m · liter)

Ø	σ	Ø	σ	Ø	σ
0.2	7032000	4.5	617	16	13.7
0.4	879000	5.0	450	17	11.5
0.6	260500	5.5	338	18	9.65
0.8	109900	6.0	260	19	8.20
1.0	56260	6.5	205	20	7.03
1,2	32560	7.0	164	22	5.28
1.4	20500	7.5	133	24	4.07
1.6	13730	8.0	110	26	3.20
1.8	9646	8.5	91.6	28	2.56
2.0	7032	9.0	77.2	30	2.08
2.2	5283	9.5	65.6	32	1.72
2.4	4070	10.0	56.3	34	1.43
2.6	3201	11	42.3	36	1.21
2.8	2563	12	32.6	38	1.04
3.0	2084	13	25.6	40	0.88
3.5	1312	14	20.5		
4.0	879	15.	16.7		

$$\sigma = \frac{10^5 \cdot 3 \cdot 6.9078}{2\pi r^3 u}. \tag{5.29}$$

The factor 10^5 is inserted to convert cgs units to meters and liters.

The mean molecular velocity can be determined from the formula

$$u = \sqrt{\frac{8RT}{\pi M}} = 1455 \sqrt{\frac{T}{M}} \quad \text{cm/sec.} \tag{5.30}$$

In the preceding table we have determined τ from Eq. (5.33), which comes from Eq. (5.25), the first of Eqs. (3.6) with $p_2 = 0$, and Eqs. (5.13), (5.28), (5.31), and (5.32):

$$p = \pi n m u^2/8, \tag{5.31}$$

$$J_{pV} = \pi u^2 m J/8, \tag{5.32}$$

$$\tau = \frac{10^3 \cdot 4 \cdot 6.9078}{WSu}. \tag{5.33}$$

We have multiplied by 10^3 in order to change from cgs units to meters. We take W from the table.

To change to other gases we must multiply by $\sqrt{\frac{M_{gas}}{M_{nitr}}}$. For example, for hydrogen we get a pumping time shorter by a factor of about four. For gas mixtures such as air, an exact calculation requires that we consider the pumping time separately for the various components, and there is a partial separation. The dependence of the pumping speed on the nature of the gas cannot in general be neglected. Times required to lower the pressure by factors other than 1000 can also be determined easily. It can be seen from Eq. (5.18) that these times are proportional to the logarithm of the pressure ratio, so that the pumping time must be multiplied by $q/3$ if the ratio is 10^q. The calculated pumping times will of course be true only when the decrease of pressure falls within the range in which the condition is valid – where the means free path is much larger than the diameter of the tube.

Let us consider one example of the application of these tables and formulas. Suppose that there is a continuous release of gas in a vessel – for example, by degassing of an electrode. If no pumping is applied, the pressure p will increase by an amount Δp per second. What must be the pumping speed so that p will remain equal to its initial small value? According to Eq. (5.17) we have the expression

$$V \cdot \Delta p = \frac{p_1}{\frac{1}{D_p} + \omega}. \tag{5.34}$$

Then from Eq. (5.18) and for $p_{10}/p_1 = 1000$ we get

$$\frac{p_1}{\Delta p} = \frac{t}{6.9078} \approx \frac{t}{7}. \tag{5.35}$$

Thus if we know the rate at which gas is released, with Eqs. (5.35) and (5.37) and by means of the table we can design a vacuum system for maintaining the desired low pressure in the vessel.

These sample calculations made by Clausing more than thirty years ago are still sufficiently good at the present time. Additional information on the construction and design of high-vacuum systems can be obtained from the papers of Tyagunov (1948); Dushman (1949); Jackel (1950); Groszkowski (1957); Yarwood (1955); and in the collection edited by Cherepnin (1963).

SECTION 6

THE MOTION OF RAREFIED GASES IN CONICAL AND BENT TUBES

A very important application of the effusion of gases through apertures lies in determining vapor pressure by the Knudsen method. Because of the difficulty of realizing an ideal effusion aperture with a Clausing coefficient equal to unity, many investigations have made use of effusion cells with conical type openings (Fig. 9). The rate of effusion from such cells is regarded as being the same as from cells with ideal effusion apertures. The lack of a calculation of the effects of the conical aperture geometry diminishes the accuracy of such vapor pressure measurements.

We shall first consider in more detail the calculation of the Clausing coefficient for a conical effusion aperture as proposed by Balson (1961).

Figure 10 shows the situation in a conical aperture with half-angle β, entrance orifice of radius r, exit orifice of radius R, and length L. The sphere is divided by the conical aperture into three zones: A, B, C.

The emergent flux consists of: (1) molecules that collide with the walls of the aperture; (2) molecules that pass through the aperture without striking the walls.

The probability of the first type of particles getting past the conical aperture is

$$W_1 = \frac{S_B}{S_B + S_C} \cdot \frac{S_C}{S_A + S_C}, \qquad (6.1)$$

where the first factor is the probability of hitting a wall and the second is the probability of leaving the wall and entering the zone C.

The probability of getting through without collision is given by the formula

$$W_2 = \frac{S_C}{S_B + S_C}. \qquad (6.2)$$

Consequently, the total probability for passing through the aperture will be given, according to the theorem of addition of probabilities, by

$$\overleftarrow{W} = W_1 + W_2 = \frac{S_C(S_A + S_B + S_C)}{(S_B + S_C)(S_A + S_C)}. \qquad (6.3)$$

where S_A, S_B, S_C are the areas of the corresponding parts of the spherical surface.

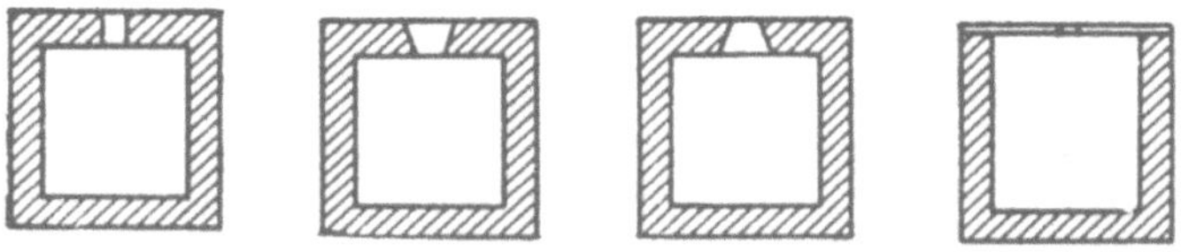

Fig. 9. Cross sectional outlines of Knudsen cells (Iczkowski, Margrave, and Robinson).

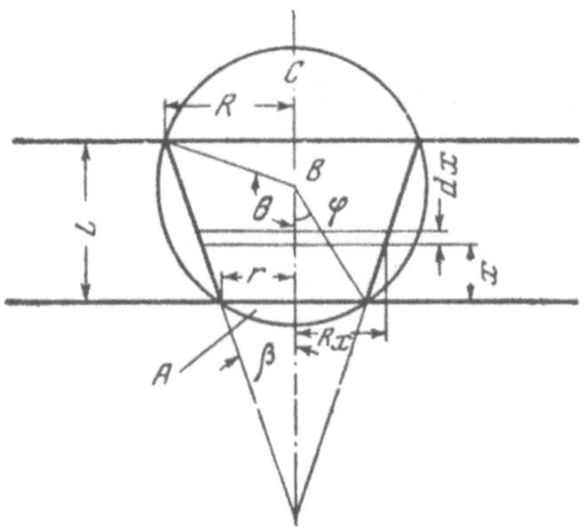

Fig. 10. Calculation of the Clausing coefficient for conical apertures (Balson).

The quantity $\overleftarrow{W}$ can be expressed in terms of the angles shown in Fig. 10 in the following way:

$$\overleftarrow{W} = \frac{(1+\cos\theta)\,2}{(1+\cos\psi)(2-\cos\psi+\cos\theta)}, \tag{6.4}$$

where

$$\cot\psi = \tan\beta + \frac{L}{2r}(1+\tan^2\beta) = \frac{R^2+L^2-r^2}{2rL},$$

$$\tan\theta = \frac{2LR}{R^2-L^2-r^2}.$$

Let us find the limits of applicability of Eqs. (6.3) and (6.4), assuming that $\beta = 45°$ (Fig. 10).

Using the Walsh formula for the disks of Fig. 10, we can write

$$w_{dd}(x) = \frac{1}{2R^2}\left[R^2+r^2+x^2-\sqrt{(R^2-x^2-r^2)^2+4x^2R^2}\,\right]. \tag{6.5}$$

It can be shown that the total probability $\overrightarrow{W}$ of a particle passing through in the opposite direction (through the convergent cone) is given by the expression

$$\overrightarrow{W} = \frac{r^2}{R^2}\left[1-\int_0^L w_{dP_x}(x)\left(1-\frac{\Phi(x)}{a}\right)dx\right]. \tag{6.6}$$

Since the rate of emission $\Phi(x)$ from a unit area at distance of x along the wall of the aperture cannot be much smaller than the corresponding value a at the entrance to the cone, the last integral of the expression (6.6) is rather small and can be determined graphically without a large error.

It can be seen from Fig. 10 that the rate of arrival $2\sqrt{2}\pi(r+x)\Phi dx$ mole/sec on a band (ring) located at the distance x is determined by the particles which arrive from the lower entrance, for the series of bands y between 0 and x and the series of bands z between x and L.

TABLE 6.1. Probability of Transmission Through a Divergent Conical Aperture (Balson)

L/r	$\overleftarrow{W}$ (6.4)	$\overleftarrow{W}$ (6.9)	W (Clausing)
0.2	0.9910	0.9902	0.9092
0.4	0.9865	0.9862	0.8341
0.6	0.9845	0.9840	0.7711
0.8	0.9838	0.9828	0.7177
1.0	0.9838	0.9783	0.6720
2.0	0.9866	0.9720	0.5136
3.0	0.9913	0.9662	0.4205
4.0	0.9937	0.9609	0.3589

Consequently

$$\frac{\Phi(x)}{a} = w_{b_x d_l}(L-x) + \int_0^x w_{P_x b_{x-y}}(y)\frac{\Phi(x-y)}{a}\,dy + \int_0^{L-x} w_{b_x d_{x+z}}(z)\frac{\Phi(x+z)}{a}\,dz. \tag{6.7}$$

TABLE 6.2. Dimensions (in mm) and Transmission Probabilities of the Effusion Apertures of the Cells Used by Searcy and McNees

L	1.85	1.10	2.00	3.71
r	0.94	1.635	2.36	2.125
R	3.39	2.855	4.59	2.125
β deg	53	48	48	0
L/r	1.97	0.67	0.85	1.74
$\overleftarrow{W}$	0.995	0.987	0.987	—
W	0.510	0.752	0.706	0.546

We assume that

$$\frac{\Phi(x)}{a} = a' + b'x + c'x^2 \tag{6.7'}$$

gives the variation of $\Phi(x)/a$ along the wall of the aperture; when we then substitute Eq. (6.7') in the integral of Eq. (6.6) and carry out some transformations we find that the term on the right differs from $\Phi(x)/a$ by a quantity E(x) which is given by

$$E(x) = k_x + a'A_x + b'B_x + c'C_x. \tag{6.8}$$

After graphical integration over the variables of the given equation at several (from four to six) points along the cone, the values of a', b', c' are calculated so as to minimize E(x) in Eq. (6.8).

When L/r = 0.8, then E(0.0) = −0.0007; E(0.2) = + 0.0052; E(0.4) = −0.0002; E(0.6) = + 0.0022; E(0.8) = −0.0009. Accordingly we see that Eq. (6.7) is capable of giving an adequate estimate of the situation. In this case a' = 0.8388; b' = 0.3755; c' = 0.2814.

We recall that the probabilities $\overrightarrow{W}$ and $\overleftarrow{W}$ for motion through the convergent and divergent cones are related to each other

$$r^2\overleftarrow{W} = R^2\overrightarrow{W}.$$

By means of Eqs. (6.6) and (6.7) we obtain

$$\overleftarrow{W} = 1 - \int_0^L w_{dP_x}(1 - a' - b'x - c'x^2)\,dx. \tag{6.9}$$

Table 6.1 shows probabilities for getting through the diverging cone (with $\beta = 45°$) obtained from Eq. (6.4) and from the more accurate expression (6.9). Values of the Clausing coefficient for a cylinder (cone angle zero) are given for comparison.

We note that the values of $\overleftarrow{W}$ remain extremely high even for comparatively large values of L/r, especially if we compare them with the values for cylindrical apertures.

It can be concluded that the simple approximate calculations of $\overleftarrow{W}$ by means of Eq. (6.4) for L/r ≈ 1 are an adequate approximation to the more exact calculations by Eq. (6.9). This agreement can be extended for larger cone angles and higher values of L/r.

TABLE 6.3. Clausing Coefficients for Conical Apertures Which Broaden Toward the Exit (From Iczkowski, Margrave, and Robinson, 1963)

β, deg	L/r						
	0.1	0.2	0.5	1.0	2.0	5.0	10.0
0	0.952399	0.909215	0.801271	0.671984	0.514231	0.310525	0.190940
1	0.954079	0.912490	0.808852	0.685401	0.536021	0.345995	0.236829
5	0.960373	0.924763	0.837261	0.735659	0.617560	0.478646	0.408600
10	0.967347	0.938350	0.868615	0.790779	0.705799	0.617242	0.580298
20	0.97865	0.96027	0.91851	0.87642	0.83704	0.80558	0.79641
30	0.98691	0.97614	0.95344	0.93338	0.91771	0.90814	0.90611
40	0.99268	0.98701	0.97619	0.96806	0.96288	0.96046	0.96008
50	0.9964	0.9939	0.9896	0.9870	0.9857	0.9852	0.9851
60	0.9986	0.9977	0.9965	0.9959	0.9957	0.9956	0.9955
70	0.9996	0.9994	0.9993	0.9992	0.9992	0.9992	0.9991
80	1.0000	1.0000	1.0000	1.0000	1.0000	1.0000	1.0000
89	1.0000	1.0000	1.0000	1.0000	1.0000	1.0000	1.0000

Although conical effusion apertures are used very often, it is rare that their dimensions are given. A pleasant exception is the work of Searcy and McNees (1953) which dealt with a determination of the vapor pressure of rhenium silicide by the Knudsen method. In the effusion cells which they used, three had conical apertures and one had a cylindrical effusion aperture. Table 6.2 presents the dimensions (quoted from Balson, 1961) of the apertures in their cells and the corresponding transmission probabilities.

The values of $\overleftarrow{W}$ are calculated from Eq. (6.4), and the values of W are obtained, according to Clausing, for the corresponding values of $L/r = 1$. We note that the work of Searcy and McNees confirms the high penetrability of the effusion apertures which they used.

It is obviously appropriate to give the name of Clausing coefficient to the probabilities of a particle passing through conical apertures, just as for the case of cylindrical apertures.

Iczkowski, Margrave, and Robinson (1963) have given a more exact solution of the problem of finding Clausing coefficients for conical apertures. The essence of their method is the replacement of the appropriate integral equation [of the type (6.7)] by a system of algebraic equations and solving them on a computer. Their results are shown in Table 6.3.

If one needs to know the Clausing coefficient $\overrightarrow{W}$ for a conical aperture which narrows toward the exit, one can use the equation

$$\overrightarrow{W} = \frac{\overleftarrow{W}}{\left(1 + \frac{L}{r}\tan\beta\right)^2}. \qquad (6.10)$$

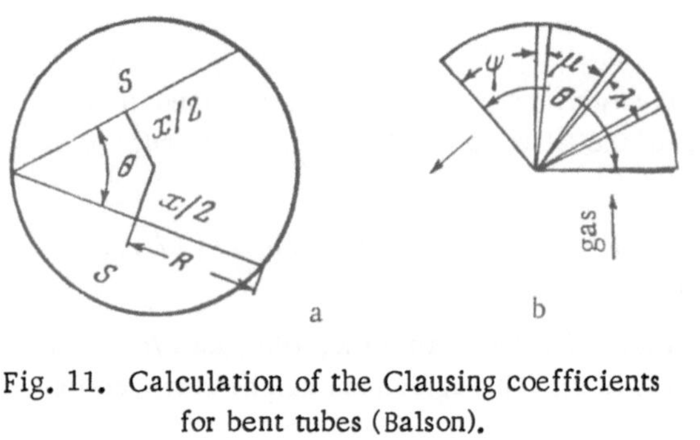

Fig. 11. Calculation of the Clausing coefficients for bent tubes (Balson).

It is a problem of unquestionable conceptual interest to find the resistance which is presented to a molecular flow by bent tubes (the calculations are due to Balson, 1961). As was pointed out in the derivation of the Walsh formulas, the relations for ω_{dd} will remain valid if the disks are not coaxial but offset, as shown for example in Fig. 11a. The disks in question are placed at an angle θ such that

$$\tan\frac{\theta}{2} = \frac{x}{2R}.$$

Then

$$w_{dd}(\theta) = \tan^2\frac{(\pi-\theta)}{4}, \tag{6.11}$$

and also

$$w_{db}(\theta) = -\frac{d}{d\theta}w_{dd}(\theta) = \frac{1}{2}\left[\tan\frac{(\pi-\theta)}{4} + \tan^3\frac{(\pi-\theta)}{4}\right]. \tag{6.12}$$

Since molecules which come from a band in which θ increases by $d\theta$ emerge from an area given by the formula $2\pi R^2 d\theta$, we have

$$w_{db}(\theta) = 2w_{bd}(\theta)\,d\theta. \tag{6.13}$$

Figure 11b shows a tube bent through a total angle θ, with bands formed by an increase $d\psi$ of the angle ψ at the exit and by two other angles λ and μ.

The rate of emergence of those molecules which never collide with the walls will be given by the expression $\pi R^2 a w_{dd}(\theta)$.

If a_ψ is the rate at which molecules leave a unit area of the band at ψ, then the rate at which they will emerge is $2\pi R^2 a_\psi w_{bd}(\psi)d\psi$.

The total efflux through the bend is

$$\int_0^\theta 2\pi R^2 w_{bd}(\psi)\, a_\psi\, d\psi = \pi R^2 \int_0^\theta w_{db}(\psi)\, d\psi.$$

The total probability of efflux is

$$W_\theta = w_{dd}(\theta) + \int_0^\theta \frac{a_\psi}{a} w_{db}(\psi)\, d\psi. \tag{6.14}$$

TABLE 6.4. Verification According to the Integral Equation (6.16) of the Applicability of Equation (6.15) (Angle of Bend $\theta = 90°$; $b = 0.41519$)

ψ, deg	0	10	20	30	40
a_ψ/a	0.2924	0.3385	0.3846	0.4307	0.4769
E_ψ	−0.0012	+0.0005	+0.0011	+0.0009	+0.0003

Thus the problem is reduced to finding how a_ϕ/a changes along the bend.

We first try a linear variation of a_ϕ/a, allowing for the fact that at 0.5θ we should get $a_\phi/a \approx 0.5$. For this reason we write

$$\frac{a_\psi}{a} = 0.5 - 0.5b + b\left(\frac{\psi}{\theta}\right). \tag{6.15}$$

The rate at which molecules leave the band of area $2\pi R^2 d\psi$ consists of three components: 1) from below, $\pi R^2 \omega_{db}(\theta-\psi)$; (2) from bands between 0 and ψ, $\int_0^\psi 2\pi R^2 a_{\psi-\mu}\, w_{bb}(\mu)\, d\mu$; and (3) from bands between ϕ and θ, $\int_0^{\theta-\psi} 2\pi R^2 a_{\psi+\lambda} w_{bb}(\lambda)\, d\lambda$. This total quantity must be set equal to $2\pi R^2 a_\phi d\phi$, so that

$$\frac{a_\psi}{a} = w_{bd}(\theta-\psi) + \int_0^\psi \frac{a_{\psi-\mu}}{a} w_{bb}(\mu)\, d\mu + \int_0^{\theta-\psi} \frac{a_{\psi+\lambda}}{a} w_{bb}(\lambda)\, d\lambda. \tag{6.16}$$

When the value of a_ϕ/a is substituted from Eq. (6.15) and the integrations are performed, it can be shown that the term on the right is larger than a_ϕ/a by the amount E_ϕ, where

$$2E_\psi = w_{bd}(\theta-\psi) - w_{bd}(\psi) + b\,[w_{bd}(\psi) - w_{bd}(\theta-\psi) + \theta^{-1}(w_{dd}(\psi) - w_{dd}(\theta-\psi))]. \tag{6.17}$$

A study of Eq. (6.17) shows that E_ψ always vanishes for $\psi = 0.5\theta$, and the values of the coefficients in Eq. (6.17) are symmetrical about $\phi = 0.5\theta$. Because of this, some five values of ϕ were chosen for angles smaller than 0.5θ. For each of the values of ψ the value of the coefficient b was taken which minimizes E_ψ. Table 6.4 shows the results of the minimization for $\theta = 90°$.

It can be seen from the data of Table 6.4 that the test equation (6.15) is entirely satisfactory, since E_ϕ neve exceeds 0.3 percent of a_ϕ/a. When we substitute the value of b calculated from Eq. (6.17) into Eq. (6.15), make use of Eq. (6.14), and perform the integration we get

$$W_\theta = \frac{1}{2}(1-b)\sec^2\left(\frac{\pi-\theta}{4}\right) + \frac{4b}{\theta}\left[1 - \tan\frac{\pi-\theta}{4}\right] - b. \tag{6.18}$$

TABLE 6.5. Clausing Coefficients for a Bend

Angle of bend θ, deg	30	60	90	120	150	180
W_0 (6.18)	0.7919	0.6497	0.5467	0.4688	0.4089	0.3623
W from Clausing	0.792	0.651	0.570	0.502	0.451	0.410

Table 6.5 gives values of W_θ obtained from Eq. (6.18). For comparison, the Clausing coefficients are given for a cylinder which has $L/r = \theta$ (θ in radians).

It can be seen from Table 6.5 that at low pressures the introduction of a bend into a tubular passage does not result in a very large effect for bends smaller than 90°; for larger angles, however, the resistance is larger than that of a cylinder of the equivalent dimensionless length.

It may be pointed out that the Balson solutions of the problem of conical apertures and of bent tubes are entirely identical. We have found approximately the same approach to the mathematical side of the problem in Chambré (1960).

SECTION 7

CALCULATION OF EFFUSION FROM A CYLINDER FOR A VAPOR OF MONOMERIC COMPOSITION

In Section 4 we derived the integral equation (4.8) for the function $\Phi(z)$, which is the lengthwise distribution along a cylinder of the number of collisions of molecules with its walls.

The direct solution of Eq. (4.8) is difficult. An indirect solution was first suggested by Buckley (1927). As can easily be shown by direct calculation, for not too large values of z the kernel $d^2F(z)/dz^2$ of the integral equation can be approximated by one or two exponentials.

Let $F(z) \cong 2e^{-z}$; then Eq. (4.8) can be written in the form

$$\Phi(z_1) = b + \frac{a}{2} e^{-z_1} + \frac{1-\beta}{2} e^{-z_1} \int_0^{z_1} \Phi(z) e^{z} dz + \\ + \frac{1-\beta}{2} e^{z_1} \int_{z_1}^{l} \Phi(z) e^{-z} dz + \frac{(1-\alpha)(1-\beta)}{2} e^{-z_1} \int_0^{l} \Phi(z) e^{-z} dz. \quad (7.1)$$

Differentiating expression (7.1) twice we get

$$\frac{d^2\Phi}{dz^2} = \beta\Phi - b, \quad (7.2)$$

$$\Phi = A \cdot e^{-z\sqrt{\beta}} + B \cdot e^{z\sqrt{\beta}} + C. \quad (7.3)$$

If we set $\beta = b = 0$, then Eqs. (7.2) and (7.3) are simplified (this case was considered by Rossman and Yarwood, 1954):

$$\frac{d^2\Phi}{dz^2} = 0, \quad (7.4)$$

$$\Phi = A_1 + B_1 z. \quad (7.5)^*$$

*One can also find in the literature on this subject other forms of $\Phi = \Phi(z)$ besides Eqs. (7.3) and (7.5). In Clausing, 1930 (9) a formula is given for long tubes:

$$\Phi(z) = \Phi(0)\left[\frac{4r}{3L} + \left(1 - \frac{8r}{3L}\right)\frac{z}{L}\right].$$

In the paper by Gunther (1957) it is assumed that the expression

$$\Phi(z) = \Phi(0)\left[\frac{4r}{4r+3L} + \left(1 - \frac{8r}{4r+3L}\right)\frac{z}{L}\right]$$

holds for arbitrary tubes (see also section 8).

After a number of transformations we find A_1 and B_1:

$$\left.\begin{aligned} A_1 &= \frac{a(l+1)}{2+\alpha l}, \\ B_1 &= -\frac{a}{2+\alpha l}. \end{aligned}\right\} \qquad (7.6)^*$$

The physical meaning of the assumption that $\beta = b = 0$ is that our solution refers to the case in which the walls do not evaporate and have a reflection coefficient of unity. In terms of radiation theory this is the case of radiation from a cylinder with perfectly white walls. It is this theory of molecular flow along a tube with walls of this sort which was given by Clausing and which many modern workers are continuing to develop.

It may be noted that the theory of Clausing relates to tubes – i.e., to cylinders without any bottom. Our solution allows us to arrive at Clausing's relations if we set $\alpha = 1$ – i.e., if we assume that there is no bottom to the cylinder or (what amounts to the same thing) that the reflection coefficient at the bottom is zero.

The solution of the more general problem without the assumption that $b = \beta = 0$ is as follows:

$$A = \frac{(1-\beta)\{b\,[(1-\alpha)(1+\sqrt{\beta}) - (1-\sqrt{\beta})]\,e^{-l\sqrt{\beta}} - (a\beta - \alpha b)(1+\sqrt{\beta})\}}{\beta\{(1-\alpha)(1-\beta) - (1+\sqrt{\beta})^2 - (1-\sqrt{\beta})[(1-\alpha)(1+\sqrt{\beta}) - (1-\sqrt{\beta})^2]\,e^{-2l\sqrt{\beta}}\}},$$

$$B = \frac{(1-\beta)\{(a\beta - \alpha b)(1-\sqrt{\beta})\,e^{-2l\sqrt{\beta}} - b[(1-\alpha)(1-\sqrt{\beta}) - (1+\sqrt{\beta})]e^{-l\sqrt{\beta}}\}}{\beta\{(1-\alpha)(1-\beta) - (1+\sqrt{\beta})^2 - (1-\sqrt{\beta})[(1-\alpha)(1+\sqrt{\beta}) - (1-\sqrt{\beta})^2]\,e^{-2l\sqrt{\beta}}\}}$$

$$C = \frac{b}{\beta}. \qquad (7.7)$$

This solution is also more general than the solution of Buckley, which is described in the literature and which can be obtained from Eq. (7.7) if we set $a = \alpha$, $b = \beta$. The physical meaning of this result is that whereas saturated black-body radiation is the same for all substances and depends only on the temperature, the saturated vapor pressure is different for different substances.

We shall write down Buckley's expressions for the coefficients $\bar{A}$, $\bar{B}$, $\bar{C}$ in Eq. (7.3) for the distribution of radiation in horizontal levels inside the cylinder. We write $\bar{A}$, $\bar{B}$, $\bar{C}$ instead of A, B, C if the radiation is measured in black-body radiation units:

$$A = \bar{A}b/\beta; \quad B = \bar{B}b/\beta; \quad C = \bar{C}b/\beta. \qquad (7.8)^{**}$$

*Owing to an inaccuracy in the calculation, Rossman and Yarwood (1954) found, instead of Eq. (7.6), expressions for A_1 and A_2 in the form

$$A_1 = \frac{a(l+1)}{2+l}; \quad B_1 = -\frac{a}{2+l}.$$

**Equations (7.7) and (7.8) define the coefficients for the equation (7.3).

Thus the problem of the distribution of monomeric flows inside a cylinder is solved for the case in which the evaporation occurs only in the form of monomeric particles.

In concluding this section we give an expression for the number of particles μ_M^b ($cm^{-2}\ sec^{-1}$) emerging from the bottom.

The integral expression is of the form

$$\mu_M^b = a - \frac{1-\alpha}{2}\int_0^l \Phi(z)\,F'\,dz, \tag{7.9}$$

since the number of particles which evaporate from the bottom per unit area and unit time is a, and a fraction $(1-\alpha)$ of the particles which strike the bottom per unit area and unit time is reflected.

If we substitute in Eq. (7.9) the values obtained earlier for F(z) and Φ(z) we get the result

$$\mu_M^b = \frac{b}{\beta} + \frac{4b\sqrt{\beta}(1-\alpha)e^{-l\beta^{1/2}} + (a\beta - \alpha b)[(1-\sqrt{\beta})^2 e^{-2l\beta^{1/2}} - (1+\sqrt{\beta})^2]}{\beta\{(1-\alpha)(1-\beta) - (1+\sqrt{\beta})^2 - [(1-\alpha)(1-\beta) - (1-\sqrt{\beta})^2]e^{-2l\beta^{1/2}}\}} \tag{7.10}$$

If we set $a = \alpha$ and $b = \beta$, we get

$$\bar{\mu}_M^b = 1 + \frac{4(1-\alpha)\sqrt{\beta}e^{-l\beta^{1/2}}}{(1-\alpha)(1-\beta) - (1+\sqrt{\beta})^2 + [(1-\sqrt{\beta})^2 - (1-\alpha)(1-\beta)]e^{-2l\beta^{1/2}}}. \tag{7.11}$$

Setting $\alpha = \beta$, we get the solution given by Buckley:

$$\bar{\mu}_M^b = 1 + \frac{4\sqrt{\alpha}(1-\alpha)e^{-l\alpha^{1/2}}}{(1-\alpha)^2 - (1+\sqrt{\alpha})^2 + [(1-\sqrt{\alpha})^2 - (1-\alpha)^2]e^{-2l\alpha^{1/2}}}. \tag{7.12}$$

Finally, for $a = \alpha \neq \beta = 0$, or (which is equivalent) for $l = 0$

$$\bar{\mu}_M^b = \alpha, \tag{7.13}$$

which is the same as the evaporation from a plane surface.

A study of Eq. (7.12), which relates to the case in which the walls and bottom of the cylinder are made of the same material with an evaporation coefficient (or emissive power) a, leads to interesting conclusions.

For $l \to \infty$ we have $\bar{\mu}_M^b = 1$.

The meaning of this is that when the cylinder is sufficiently long the vapor can reach saturation over the bottom.* In the terminology of radiation theory this means that the bottom of the cylinder becomes a black body (to arbitrary accuracy) (see below the results of De Vos).

*Strictly speaking, we have no right to extrapolate l to ∞, since as this is done our approximation becomes worse. If, however, we do not postulate that Knudsen's law is satisfied, but regard the approximation we have taken for the kernel as exact, these objections have no force [the same comment applies also to Eqs. (7.17), (7.25), and so on].

We are led to somewhat different conclusions by an analysis of the more general expressions (7.10) and (7.11), since neither of these formulas gives saturation over the bottom for $l\to\infty$. Let us consider, for example, the value for μ_M^b which will be obtained from Eq. (7.10). For $l\to\infty$ we get:

$$\mu_M^b(\infty)=\frac{b}{\beta}-\frac{(a\beta-\alpha b)(1+\sqrt{\beta})^2}{3[(1-\alpha)(1-\beta)-(1+\sqrt{\beta})^2]} \qquad (7.14)$$

The value of μ_M^b given by Eq. (7.14) depends in a rather complicated way on α, β, a, and b, and saturation over the bottom of the cylinder becomes problematical.

We present the solution of the problem of the rate of evaporation from the bottom of the cylinder on the assumption that $b=\beta=0$.

The solution can be carried out either in the same way as the preceding one, that is, by substituting into the integral expression (7.9) the equation for $\Phi(z)$ in the case $b=\beta=0$, Eq. (7.5), or else by calculating the limit of the expression (7.11) for $b\to 0$ and $\beta\to 0$.

Both procedures lead to the same result

$$\mu_M^b=\frac{2a+al}{2+\alpha l}. \qquad (7.15)$$

The special (limiting) cases are obviously not open to any objection:

1) for $l=0$ $\quad\mu_M^b=a,$ (7.16)

2) for $l\to\infty$ $\quad\mu_M^b=\frac{a}{\alpha},$ (7.17)

3) for $\alpha\to 1$ $\quad\mu_M^b=a.$ (7.18)

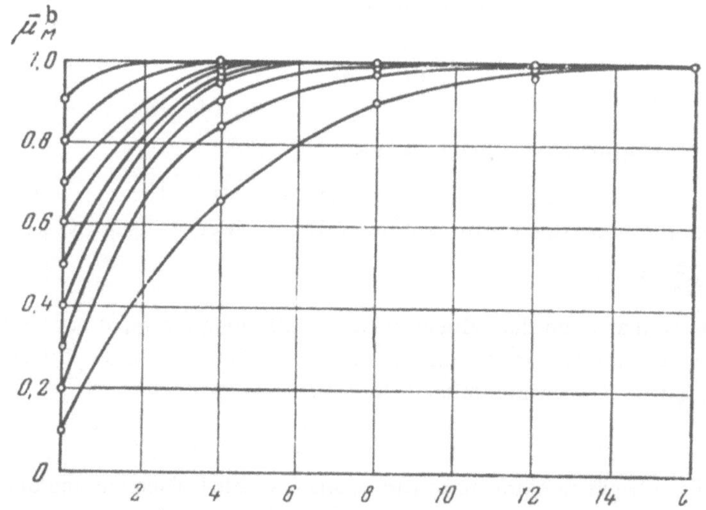

Fig. 12. Degree of saturation of the vapor near the bottom of the cylinder as a function of cylinder length and of the evaporation coefficient [Eq. (7.12)].

The expressions obtained are of some interest for the analysis of the behavior of a cylinder bottom from which evaporation takes place. Figure 12 is supplied as a demonstration of the results. As can be seen, for sufficiently large sizes the bottom "forgets" about the evaporation coefficient, which has an influence only for small sizes.

Equations for the Total Flow of a Monomer from a Cylinder

Particles pass from the walls and the bottom of the cylinder to a receiver outside the cylinder. Along the z-axis of the cylinder there is a distribution of flux $\Phi(z)$. The particles leaving the bottom of the cylinder are both those which have just been evaporated and those which are reflected after having come from the walls. Besides this, both the bottom and the walls reflect particles onto each other without thus making any contribution to the flux of particles to the outside of the cylinder.

We consider each of the indicated portions of the flux, and then add them together. According to the formula of Walsh, the number of particles which come to the receiver from the bottom without touching the walls of the cylinder is $aF(l)/2$ particles per square centimeter per second. The number of particles which arrive at the bottom from the walls is given by the formula

$$-\frac{1}{2}\int_0^l \Phi(z)\,F'\,dz.$$

Of these particles a fraction $(1-\alpha)$ is reflected.

A certain fraction of the reflected particles again goes to the walls, as has been taken into account in the derivation of the equation for the distribution of the outgoing particles along the length of the cylinder. The number striking per unit area of the exit aperture is

$$-\frac{1-\alpha}{4}F(l)\int_0^l \Phi(z)\,F'dz \text{ particles (cm}^{-2}\text{ sec}^{-1}). \tag{7.19}$$

Accordingly, the number of particles per square centimeter per second coming from the bottom and arriving at the exit aperture is

$$F(l)\left[\frac{1}{2}a-\frac{1-\alpha}{4}\int_0^l \Phi F'dz\right],$$

and the number coming from the walls to the exit aperture is

$$\frac{1}{2}\int_0^l \Phi(z)\,F'(l-z)\,dz.$$

Thus the total number of particles which arrive per unit area of the receiver per unit time, or the rate of effusion μ of particles from the cylinder, will be

$$\mu=\frac{1}{2}aF(l)-\frac{1-\alpha}{4}F(l)\int_0^l \Phi(z)\,F'\,dz+\frac{1}{2}\int_0^l \Phi(z)\,F'(l-z)\,dz. \tag{7.20}$$

For $F(l) \cong 2e^{-l}$

$$\mu = ae^{-l} + (1-\alpha)e^{-l}\int_0^l \Phi(z)e^{-z}dz + e^{-l}\int_0^l \Phi(z)e^{z}dz, \tag{7.21}$$

where $\Phi(z)$ can be determined from the formulas (7.3), (7.5)–(7.7).

It is clear that by taking different particular expressions for Φ one can also get different values for μ, depending on the special conditions of the problem.

We shall solve the problem in the most general form, and then examine special cases. In the last expression, Eq. (7.21), we substitute the value of Φ from Eq. (7.3) with the coefficient (7.5), and after some calculations we get

$$\mu = \frac{b\left[(2\alpha\beta-4\beta-2\alpha\sqrt{\beta})+(4\beta-2\alpha\beta-2\alpha\sqrt{\beta})e^{-2l\sqrt{\beta}}+4\frac{\alpha}{b}\sqrt{\beta}(\alpha b-a\beta)e^{-l\sqrt{\beta}}\right]}{\beta\{(1-\alpha)(1-\beta)-(1+\sqrt{\beta})^2-[(1-\alpha)(1-\beta)-(1-\sqrt{\beta})^2]e^{-2l\sqrt{\beta}}} \tag{7.22}$$

If we set $a=\alpha$, $b=\beta$, or, what amounts to the same thing, substitute in Eq. (7.21) the expression (7.3) with the coefficients (7.8), we then get from Eq. (7.22):

$$\bar{\mu} = \frac{2\alpha\beta-4\beta-2\alpha\sqrt{\beta}+(4\beta-2\alpha\beta-2\alpha\sqrt{\beta})e^{-2l\sqrt{\beta}}}{(1-\alpha)(1-\beta)-(1+\sqrt{\beta})^2+[(1-\alpha)(1-\beta)-(1-\sqrt{\beta})^2]e^{-2l\sqrt{\beta}}}. \tag{7.23}$$

If the bottom and the walls of the cylinder are made from the same material, we can set $\alpha=\beta$, and

$$\bar{\mu} = \frac{2\alpha^2-4\alpha-2\alpha\sqrt{\alpha}+(4\alpha-2\alpha^2-2\alpha\sqrt{\alpha})e^{-2l\sqrt{\alpha}}}{\alpha^2-3\alpha-2\sqrt{\alpha}+[3\alpha-2\sqrt{\alpha}-\alpha^2]e^{-2l\sqrt{\alpha}}}. \tag{7.24}$$

For $l\to 0$ we have $\mu=\alpha$, and for $l\to\infty$

$$\bar{\mu}(\infty) = 1+\frac{(\alpha-1)(\alpha-2\sqrt{\alpha})}{(\alpha^2-3\alpha-2\sqrt{\alpha})}. \tag{7.25}$$

The second term on the right-hand side of Eq. (7.25) is negative and is not equal to 0 if $\alpha \neq 1$. Consequently the flow of vapor even from an infinitely long cylinder is not a saturated vapor flow (or the total radiation from such a cylinder is not black) if α is not too large.

By calculation from Eq. (7.25) we can get a notion of the degree of saturation of the vapor for different values of α.

α	0.1	0.5	0.9
$\bar{\mu}(\infty)$	0.48	0.83	0.99

The investigation of Eq. (7.22) is somewhat more complicated, but does not give essentially different results.

It is seen that there can be evaporators which do not give an effusion coefficient $\bar{\mu}$ equal to unity [Eq. (7.25)], even though there is still a definite part of the evaporator (over the bottom) where the vapor is saturated [see Eq. (7.12) and the subsequent discussion].

As has been pointed out, Clausing assumes that the particles of a gas are completely and diffusely reflected from the walls of the bounding vessel and that there is absolutely no adsorption, to say nothing of chemisorption or chemical reactions at the surfaces.

Let us find the Clausing coefficient W_α with the same assumptions, but for a cylinder made of a material for which the reflection coefficient is $(1-\alpha)$.

We can get the solution either by substituting Eq. (7.5) with the coefficients (7.6) in the expression (7.21), or else by substituting in the general expression (7.22) the values $\beta = b = 0$. Both methods lead to the same results:

$$\mu = \frac{2a}{2+\alpha l}, \tag{7.26}$$

from which the desired Clausing coefficient is found to be

$$W_\alpha = \frac{1}{1+0.5\alpha l}. \tag{7.27}$$

The formula (7.26) was found in 1954 by Rossman and Yarwood by the method described here.

The same results, but in the special case $\alpha = 1$, were obtained empirically by Kennard (cf. Dushman, 1950), and also quite recently by V. I. Lozgachev (1963).*

In all of the calculations we have given it was assumed that there is no specular reflection at all, whereas actually this sort of reflection plays some part (not a large one). A paper by De Vos (1954) shows a method for including the contribution of the specular reflection superposed on the diffuse reflection. We shall also give the method of De Vos, for the reason that it evidently applies also to the most general case of inelastic interactions of molecular flows with the surfaces which bound them. The theory of these interactions is in its beginning stages.

The theory of De Vos has been developed only in the terminology of photometry, but this need not discourage us from using it for the calculation of molecular flows as well.

It has been seen that in almost all of the calculations we assume monochromatic (for photometry) or monoenergetic (for molecular flows) quantities. This also applies to the theory of De Vos.

De Vos introduces, in addition to the usual hemispherical reflecting power ρ^A, the partial reflecting power r^{AN}.

The first of these gives the fraction of those particles coming from A which are reflected into the hemisphere, while the second gives the fraction of particles arriving from the same direction but reflected in the direction of N.

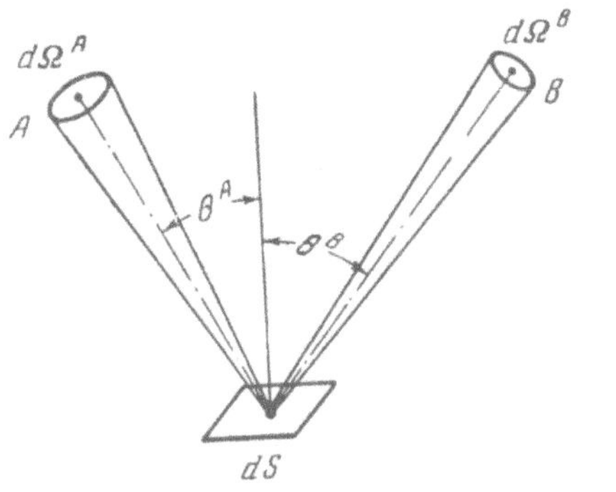

Fig. 13. Illustration of the concepts of De Vos.

The relation between ρ^A and r^{AN} is defined by the equation

$$\rho^A = \int_{2\pi} r^{AN} d\Omega^N, \tag{7.28}$$

where $d\Omega^N$ is an elementary solid angle in the direction of N.

*In Lozgachev's notation, which we use in Section 9, the symbol $\omega(1)$ appears instead of W, but this does not change the essential point.

The reciprocity theorem (cf. Fig. 13) yields

$$r^{AB}\cos\theta^{A} = r^{BA}\cos\theta^{B}, \qquad (7.29)$$

It is possible to conceive of an analog to Eq. (7.28) if we assume that a part of the radiation incident on the hemisphere is reflected in a particular direction.*

We can introduce the directional reflecting power ρ_N, defining it as that fraction of the particle flux arriving from all directions which is reflected in a given direction N (dropping N from the super- to the subscript position). Our analog of Eq. (7.28) is

$$\rho_N = \int_{2\pi} r^{AN}\, d\Omega_1^A, \qquad (7.28')$$

where $d\Omega^A$ is an elementary solid angle in the direction from whence the radiation arrives.

If ρ^A does not depend on the direction of incidence and is equal to the reflecting power at normal incidence $\rho^{\perp}$, then

$$r^{AN} = \frac{\rho^{\perp}\cos\theta^{N}}{\pi}, \qquad (7.30)$$

When displayed graphically, the scattering characteristic for perfectly diffuse reflection will represent a sphere tangent to the reflecting area, with a diameter $\rho^{\perp}/\pi$.

If the law (7.30) is not obeyed, lobes or indentations will appear on the surface of the sphere, corresponding to departures from diffuse reflection.

Examples of deviations from diffusivity obtained by De Vos with a constant $\rho^{\perp} = 0.6$ are shown in Fig. 14 (a to d).

The arrows in the figures indicate the directions of incidence. The tips of the arrows point to the reflection curve due to radiation impinging in the direction of the given arrow.

The values of the partial reflecting power are given on a logarithmic scale.

We now present the computational procedure proposed by De Vos. Let the radiant emittance per unit wavelength interval, denoted by I, be given by the Planck or Wien radiation laws:

$$I = C_1 \cdot \lambda_c^{-5}\, e^{-C_2/\lambda_c T},$$

where $C_1 = 8\pi ch$, $C_2 = ch/k_B$, c is the velocity of light, h is the Planck constant, k_B is the Boltzmann constant, and λ_c is the optical wavelength.

The radiant emittance of any element of the surface is made up of the self-radiation, plus the reflection of the radiation arriving from other elements.

*Such a situation arises in cathode sputtering (Wehner, 1959, 1960) and in connection with the growth of crystals from the vapor phase (Volmer, 1921).

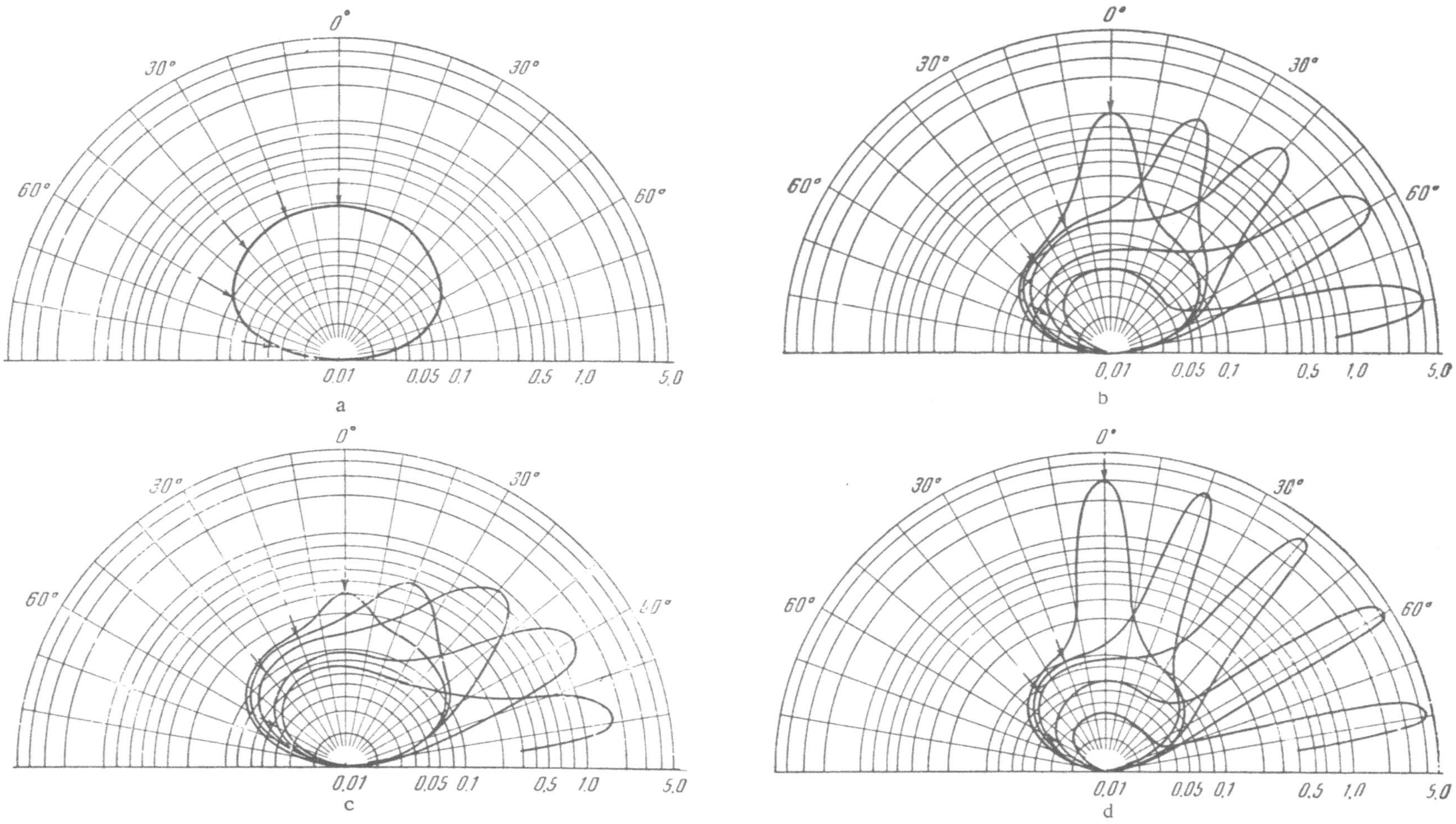

Fig. 14. Reflection characteristics of a surface for various percentages of specular reflection (De Vos).

For the calculations we use the model of an ideal black body of fairly arbitrary configuration with one aperture (Fig. 15).

We calculate the radiation intensity I_W^O transmitted by some element dW through an opening dO, taking into account reflected radiation arriving from the remaining surface elements dN of the cavity. The calculation is carried out by the method of successive approximations.

We first assume that $I_W^O = I$, whereupon the amount of radiation in the direction θ_W^O inside the solid angle $d\Omega_W^O$ per unit time per unit wavelength interval is determined by the expression

$$\varepsilon_W^O I dW \cos\theta_W^O \, d\Omega_W^O, \tag{7.31}$$

where ε_W^O is the emittance of dW at temperature T in the direction θ_W^O, and $d\Omega_W^O$ is the solid angle subtended by dO and dW. The amount of radiation which comes from the given surface element dN and is reflected by dW into the solid angle $d\Omega_W^O$ is given by the expression

$$I_N^W \, d\Omega_W^N \, dW \cos\theta_W^N r_W^{NO} \, d\Omega_W^O,$$

where I_N^W is the radiant emittance of dN in the direction of dW at temperature T, $d\Omega_W^N$ is the solid angle subtended by dN as seen from dW, and r_W^{NO} is the partial reflecting power of dW at temperature T in the direction from dN to dW and then on to dO.

The partial reflecting powers r_W^{NO} must be taken for all directions in the hemisphere by means of the reflection diagrams of the element dW. This procedure can be simplified, however, by using Eq. (7.29):

$$\cos\theta_W^N r_W^{NO} = \cos\theta_W^O r_W^{ON},$$

and the integral reduces to the expression

$$dW \cos\theta_W^O \, d\Omega_W^O \int I_N^W r_W^{ON} \, d\Omega_W^N. \tag{7.32}$$

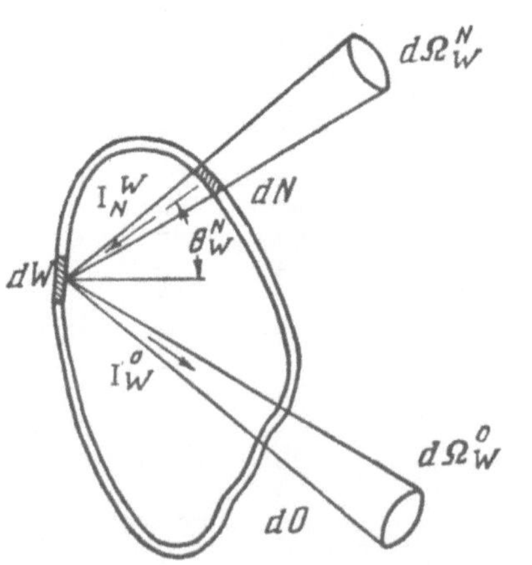

Fig. 15. Model of a black body of arbitrary shape with one aperture (De Vos).

The values of r_W^{NO} can be taken from one of the reflection diagrams (Fig. 14) with the angle of incidence θ_W^O:

$$\int r_W^{ON} \, d\Omega_W^N = \int_{2\pi} r_W^{ON} \, d\Omega_W^N - r_W^{OO} \, d\Omega_W^O,$$

and, according to Eq. (7.28), this is equal to

$$\rho_W^O - r_W^{OO} \, d\Omega_W^O.$$

Thus, the amount of radiation from all of the elements of the surface which is reflected by dW into the solid angle $d\Omega_W^O$ is

$$I dW \cos\theta_W^O \, d\Omega_W^O \left(\rho_W^O - r_W^{OO} \, d\Omega_W^O\right).$$

Adding to this quantity the radiation given by Eq. (7.31), we get the total radiation sent out into $d\Omega_W^O$.

If we disregard the fact that a portion of the thermal radiation passes through the substance, we have on the basis of the law of conservation of energy

$$\rho_W^O + \varepsilon_W^O = 1,$$

and the aperture in our black-body model will be irradiated as

$$I_W^O = I\,(1 - r_W^{OO}\, d\Omega_W^O).$$

Additional slit openings (if such exist) may be accounted for by adding terms of the type $r_O^{ON}\, d\Omega_W^N$ to the factor in the parentheses:

$$I_W^O = I\left(1 - \sum_N r_W^{ON}\, d\Omega_W^N\right) = I\varepsilon_o. \tag{7.33}$$

The factor in the parentheses, denoted by ε_O, may be called the aperture emittance of the model. Equation (7.33) gives the first approximation. To obtain the next approximation we assume that the surface radiates according to Eq. (7.33) rather than according to the Wien law.

Let the surface element dN have a temperature T_n, so that the flux I_N^W from dN onto dW will be defined in the first approximation as

$$I_N^W = I\,(1 - \sum_N r_N^{WN}\, d\Omega_W^N - k_N\, \varepsilon_N^W), \tag{7.34}$$

where ε_N^W is the emittance of dN in the direction of dW, and

$$k_N = \frac{I\,(T) - I\,(T_N)}{I\,(T)}\,.$$

If the difference $(T - T_N)$ is small, k_N may be calculated by differentiating the Wien law with respect to T:

$$k_N = \frac{c_2 \Delta T}{\lambda_c T^2}\,. \tag{7.35}$$

The resultant radiation equation takes the following form in the second approximation:

$$I_W^O = I\left(1 - \sum_i r_W^{Oi}\, d\Omega_W^i - \sum_i \int r_N^{Wi}\, d\Omega_N^i r_W^{ON}\, d\Omega_W^N - \int k_N \varepsilon_N^W\; r_W^{ON}\; d\Omega_W^W\right). \tag{7.36}$$

In the above expression the integration is carried out over the entire surface except for the opening.

De Vos notes that with the application of well-constructed black-body models the influence of the third approximation is no longer appreciable, and when the model is of low quality there is no point in performing a precise calculation anyway. This is only true, of course, if we are concerned with black-body models, as opposed, for example, to ascertaining the radiative losses of a body, etc.

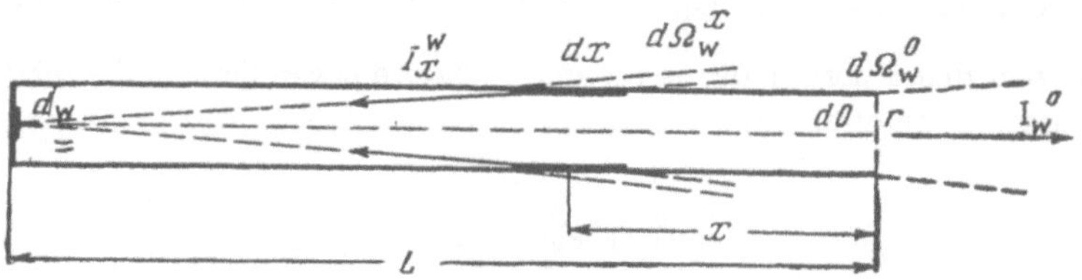

Fig. 16. Cylindrical black body closed at one end (De Vos).

TABLE 7.1. Dependence of the Quality ε_0 of a Cylindrical Black Body in the First (I) and Second (II) Approximations for Different Values of l

l	Fig. 14a		Fig. 14b		Fig. 14c		Fig. 14d	
	I	II	I	II	I	II	I	II
6	0.983	0.970	0.977	0.957	0.948	0.900	0.913	0.833
10	0.994	0.990	0.992	0.986	0.978	0.962	0.945	0.900
15	0.997	0.995	0.996	0.994	0.989	0.983	0.973	0.961
20	0.998	0.997	0.998	0.997	0.993	0.990	0.983	0.978
30	0.999	0.999	0.999	0.999	0.996	0.995	0.992	0.990

To conclude this section we give a sample calculation of the quality of a cylindrical black body by the De Vos method (see Fig. 16).

We assume that the temperature gradient is negligibly small so that we can omit the last term of Eq. (7.36). In first approximation [Eq. (7.33)]

$$I_W^0 = I\left(1 - r_W^{00}\frac{\pi r^2}{L^2}\right). \tag{7.37}$$

Introducing the dimensionless length $l = L/r$, we get

$$\varepsilon_0 = 1 - r_W^{00}\frac{\pi}{l^2}.$$

We now give the formula for ε_0 in the second approximation. Let dN be a band element of length dx; then the respective formulas for the solid angles are

$$d\Omega_W^x = \frac{2\pi r^2\, dx}{[(L-x)^2 + r^2]^{3/2}}; \quad d\Omega_x^0 = \frac{\pi r^2}{x^2 + r^2}.$$

When $L/r = l$ and $x/r = z$, we get

$$\varepsilon_0 = 1 - r_W^{00}\frac{\pi}{l^2} - 2\pi^2\int_0^l \frac{r_W^{0z} r_z^{W0}\, dz}{(z^2+1)\,[(z-l)^2+1]^{3/2}}. \tag{7.38}$$

Integrating Eq. (7.38) numerically, we obtain Table 7.1, which illustrates the remarks of De Vos presented above.

An analysis of Table 7.1 leads to the natural conclusion that increasing the fraction of specular reflection (from Fig. 14a to d), all other conditions remaining the same, decreases the quality of the black body. Consequently, the true picture is distorted to the same degree, as is our notion of completely diffuse scattering of molecules at the walls and emission of molecules from the walls of the vessel.

De Vos (1954) also gives the calculations for other types of black-body models, but we omit these data. At the risk of being repetitious, we believe that ε_0 corresponds to the effusion coefficient $\bar{\mu}$.

SECTION 8

CALCULATION OF THE DIRECTIVITY DIAGRAMS FOR MOLECULAR FLOW FROM A CYLINDER

All of the investigations of which we are aware relating to the problem have been conducted on the hypothesis that the area of the molecular source is considerably smaller than the area of the detector, which is equivalent to saying that the molecular source is a point source. Unfortunately, these conditions are usually taken for granted without being explicitly stated as such. This restriction is upheld in the present section.

Let the z axis be some fixed direction, for example the axis of a cylindrical source of molecular flow (Fig. 17), and let θ be the angle between the z axis and a chosen variable direction. The directivity diagram (or the directionality indicatrix) shows the dependence of the flux intensity I of the molecules on the angle θ. It is often convenient to express this dependence in the form

$$\frac{I}{I_0} = f(\theta),$$

where $I_0 = I$ for $\theta = 0$.

Clausing (1930) (9) was the first to show that for a cylindrical point source of molecular flow the problem of finding $f(\theta)$ is closely connected with that of finding the function $\Phi(z)$ (see Sec. 7)— the density distribution, along the length of the source, of collisions of particles with its walls. Clausing calculated the directivity diagram for molecules flowing out through a short cylinder into a vacuum. The solution was found on the assumption that there is completely diffuse scattering at the walls of the channel and that the mean free path is larger than the dimension of the cylinder. Clausing's calculation is for a cylinder without a bottom (Fig. 17).

For the case $L \to 0$ the distribution of flow of the molecules from the vessel I into the vacuum II would follow the cosine law (see Sec. 1). It was found by Clausing [1926, 1928, 1930 (5)] that for short tubes the function $\Phi(z)$ is given by the formula* [cf. Eq. (5.12)]:

$$\Phi(z) = a\left[\gamma + (1 - 2\gamma)\frac{L - z}{L}\right], \tag{8.1}$$

$$\gamma = \frac{\sqrt{L^2 + 4r^2} - L}{2r + \dfrac{4r^2}{\sqrt{L^2 + 4r^2}}}; \tag{8.2}$$

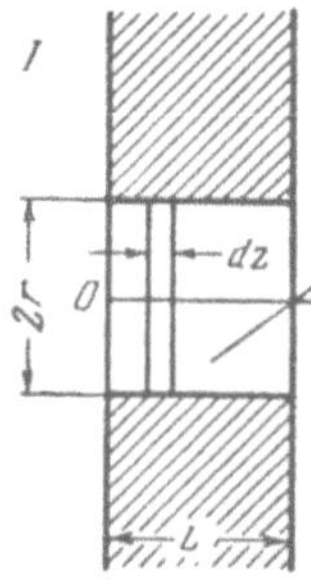

Fig. 17. A short cylindrical channel (Clausing).

Here a is the number of molecules which are incident per square centimeter and per second in the first reservoir.

Somewhat anticipating our later discussion, we note the demonstration by Ivanov and Troitskii (1963), applying modern computing techniques, that the function $\Phi = \Phi(z)$ is linear for $l = 1-80$:

$$\Phi(z) = a[k_0 + k_1(L - z)], \tag{8.3}$$

*For other solutions for $\Phi(z)$ see Sec. 7.

where

$$k_0 = \frac{1 + 2l^{-2}}{1 + (l + 2)\, l^{-2}} + \frac{(2 + l^2)(4 + l^2)^{-\frac{1}{2}} - l - 1}{1 + 2\,(4 + l^2)^{-\frac{1}{2}}},$$

$$k_1 = \frac{1 + L - (2 + l^2)(4 + l^2)^{-\frac{1}{2}}}{L\left[1 + 2\,(4 + l^2)^{-\frac{1}{2}}\right]}.$$

It is clear that the notation of Ivanov and Troitskii is related to that of Clausing by the formulas

$$k_0 = \gamma, \qquad k_1 = \frac{1 - 2\gamma}{L}.$$

In determining the number of molecules which emerge from reservoir I into II in the direction θ, it is helpful to distinguish two cases:

1) Molecules which come through the entrance opening of the channel and do not touch the walls contribute to the total flow:

$$\tan\theta < \frac{2r}{L}; \tag{8.4}$$

2) Molecules which come through the entrance opening and do not touch the walls do not contribute to the total flow:

$$\tan\theta > \frac{2r}{L}. \tag{8.5}$$

Clausing, 1930 (9) found the function R(s) defining the fraction of the particles which passes without colliding with the walls and in the direction θ through two cross sections of the channel characterized by the parameter s:

$$s = \frac{(L - z)\tan\theta}{2}, \tag{8.6}$$

$$R(s) = 2r^2 \arccos\frac{s}{r} - 2s\sqrt{r^2 - s^2}. \tag{8.7}$$

The fraction of the particles emerging in the direction θ from a band of width dz is found by differentiating Eq. (8.7) with respect to s, and is

$$\frac{dR(s)}{ds} = -4\,(r^2 - s^2)^{1/2}. \tag{8.8}$$

For the total number of molecules I(θ) (per unit solid angle and per second) which emerge in the direction θ we have for the case (8.4):

$$I(\theta) = \frac{\cos\theta}{\pi}\left[\int\limits_{\frac{L\tan\theta}{2}}^{0} \Phi(s)\left(\frac{dR(s)}{ds}ds + \mu_M^b R(s)_{x=0}\right)\right], \tag{8.9}$$

where, for the problem solved by Clausing (a short cylinder without bottom and with reflection coefficient at the walls equal to 1),

$$\Phi(s) = a\left\{\gamma + (1-2\gamma)\frac{2s}{\tan\theta}\right\}, \tag{8.10}$$

$$\mu_M^b = \frac{a}{\pi}, \tag{8.11}$$

$$R(s) = 2r^2\arccos\frac{L\tan\theta}{2r} - L\tan\theta\sqrt{r^2 - \frac{L^2\tan^2\theta}{4}}. \tag{8.12}$$

A simple and direct calculation with Eq. (8.8) leads to the expression

$$I(\theta) = \frac{a}{\pi}\cos\theta\pi r^2\left[1 - \frac{2}{\pi}(1-\gamma)\left(\arcsin p' + p'\sqrt{1-p'^2}\right) + \frac{4}{3\pi}(1-2\gamma)\frac{1-(1-p'^2)^{3/2}}{p'}\right] \tag{8.13}$$

$$\text{for} \quad p' = \frac{L\tan\theta}{2r} \leqslant 1.$$

For the case (8.5)

$$I(\theta) = \frac{\cos\theta}{\pi}\int\limits_0^r 4\sqrt{r^2 - s^2}\,\Phi(s)\,ds. \tag{8.14}$$

Using Eq. (8.10) and calculating, we get

$$I(\theta) = \frac{a}{\pi}\cos\theta\pi r^2\left[\gamma + \frac{4}{3\pi}\frac{1-2\gamma}{p'}\right] \tag{8.15}$$

for

$$p' = \frac{L\tan\theta}{2r} \geqslant 1.$$

Let the expressions in the brackets in Eqs. (8.13) and (8.15) be denoted by B. For the special case of a cylinder which has $L = 2r$, Clausing (1930)(9) calculated the following angular dependence $B = B_\theta$ (θ in degrees):

θ	B	θ	B	θ	B
0	1.0000	35	0.5835	70	0.3221
5	0.9444	40	0.5183	75	0.3011
10	0.8882	45	0.4611	80	0.2811
15	0.8310	50	0.4259	85	0.2617
20	0.7721	55	0.3956	90	0.2426
25	0.7114	60	0.3687		
30	0.6483	65	0.3445		

When we calculate the total flux emerging from the cylinder we obviously must get the same results as in Sections 5 and 7. In fact, we get for the total flux

$$J=\int_{\Omega} I(\theta)\frac{d\Omega}{\pi}=\pi r^2 a\iint B\cos\frac{d\Omega}{\pi}=\pi r^2 a\int_0^{\pi/2} B\sin 2\theta\, d\theta. \tag{8.16}$$

Using the data given above ($B = B_\theta$) and integrating with Simpson's rule for θ = 0, 5, 15, 30, 45, 75, 90°, Clausing found

$$J=0.512\cdot\pi r^2 a, \tag{8.17}$$

which is in good agreement with the Clausing coefficient calculated in Section 5 by a different method, which gave (for $L/r = 2$), W = 0.5136.

At angles of 5°, 10°, and 15° we note deviations from the cosine law by 6, 11, and 17%, respectively. The deviations increase with increasing angle.

In more recent times the distribution of molecular flows in space has been studied by M. N. Korsunskii and S. A. Vekshinskii (1945), who obtained approximate expressions for the directivity diagrams and checked them with silver vapor. There is also the well known work of Gunther (1957) who determined the directivity diagrams by the method of deposition of SiO_2 vapor at angles θ = 0–45° over a wide range of pressures for source channels $l \approx 0-5$.

Detkov, 1960 (3) considered the Clausing problem and found equations which agree with Clausing's solutions (8.13) and (8.15). By numerical calculations he determined the probability W(β) of a molecule entering a cone with vertex half-angle β = 15 and 30° for various values of L/2r for the case $L/2r < \cot\beta$.

Detkov's results give the following dependence:

L/2r	W (β = 15°)	W (β = 30°)
0.00	0.06699	0.2500
0.1	0.06569	0.23765
0.2	0.06462	0.23052
0.3	0.06377	0.22301
0.5	0.06192	0.21127
1.0	0.05952	0.19370

A paper by Troitskii (1961) gives the solution of the problem of the directivity diagram for sources in the form of channels with rectangular cross sections, for a wide range of pressures (from the Knudsen type of flow to Poiseuille flow). Naumov (1963) confirmed Troitskii's results in experiments with ammonia, using multichannel sources.

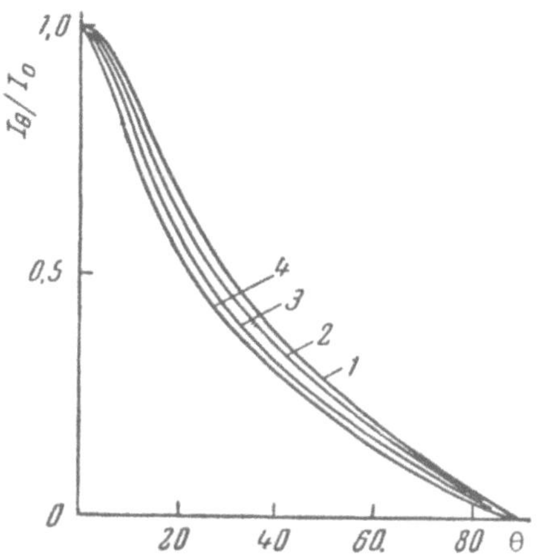

Fig. 18. Directivity diagrams of a source with L = 1 mm, 2r/L = 0.35, and diameter of the set of channels 3 mm, at various pressures (Naumov). λ = (1) 0.28 mm, (2) 0.49 mm, (3) 1.3 mm, (4) 3.4 mm.

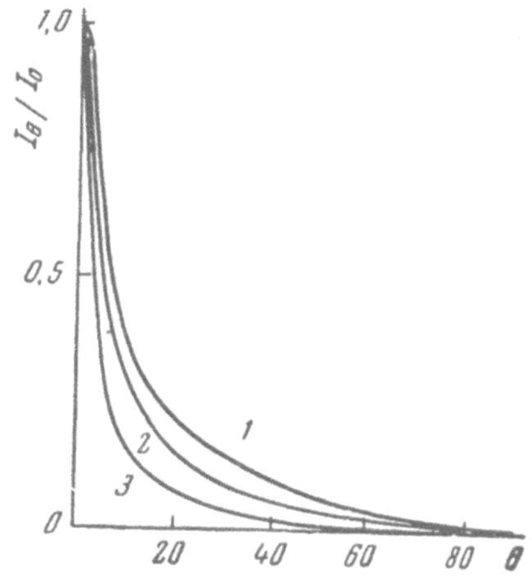

Fig. 19. Directivity diagrams of a source with L = 11.9 mm, 2r/L = 0.02, and diameter of the set of channels 6.3 mm, at various pressures (Naumov). λ = (1) 0.17 mm, (2) 0.34 mm, (3) 3.4 mm.

The results of Naumov have been shown in Fig. 3; to supplement this diagram we show also Figs. 18 and 19.

From his experimental researches Naumov came to the conclusion that in the case of molecular flow the halfwidth $\Delta\theta_{0.5}$ of the directivity diagram* is determined only by the geometry of the channel, whereas under transition conditions (from Knudsen flow to Poiseuille flow) it is a function of the total molecular flux in the beam and is determined by a certain effective length of the channel. According to the data of Naumov, for Knudsen conditions

$$\Delta\theta_{0.5} = 2\operatorname{arc\,tan}\frac{2r}{L}, \tag{8.18}$$

which is also in agreement with the theoretical results of Troitskii.

A paper by Troitskii (1961) takes into account (when there is departure from Knudsen's conditions) the emission from the volume of gas which is in the cylinder, and also the screening by this volume of the emission from the walls and the bottom. Then, if the number of particles striking the area dS of the wall per square centimeter per second is $\Phi(z)$, at the exit from the tube the flux from dS (wall) will be

$$dI_w(\theta) = \Phi(z)\,dS\,\frac{\sin\theta}{\pi}\exp\left(-\int_0^z \frac{dz}{\lambda(z)\cos\theta}\right)d\Omega,$$

where $\lambda(z)$ is the mean free path in the vicinity of z.

*The halfwidth of the directivity diagram is what Troitskii and Naumov call the angle at which the amount of matter effusing is half of that along the normal to the source.

The corresponding fluxes $dI_b(\theta)$ from the bottom and $dI_V(\theta)$ from the volume are

$$dI_b(\theta) = a dS \frac{\cos\theta}{\pi} \exp\left(-\int_0^L \frac{dz}{\lambda(z)\cos\theta}\right) d\Omega$$

and

$$dI_V(\theta) = \frac{dV\, d\Omega}{4\pi} \nu_V(z)\, \psi(u, \theta, z, y) \exp\left(-\int_0^z \frac{dz}{\lambda(z)\cos\theta}\right),$$

where $\nu_V(z)$ is the number of collisions in the volume in the vicinity of z and ψ is the distribution function with respect to the direction of flow of colliding molecules. As stated, Troitskii solved the problem for tubes of rectangular cross section.

Let us consider the problem of the directivity diagram of the molecular flow effusing from a cylinder with a bottom, in the simplified case when $b = \beta = 0$, but $\alpha \neq 0$ (see Sec. 7).

For this case

$$\Phi(z) = A_1 + B_1 z = \frac{a}{2+\alpha l}(l - z + 1).$$

By means of Eq. (8.6) we can express $\Phi(z)$ as a function of s:

$$\Phi(s) = \frac{a}{2+\alpha l}\left(\frac{2s}{\tan\theta} + 1\right). \tag{8.19}$$

Repeating all of Clausing's calculations, we get:
1) for $p' \leq 1$

$$I(\theta) = \frac{a\cos\theta}{2+\alpha l}\left\{2 + l + \frac{4l}{3\pi p'}\left[1 - (\sqrt{1-p'^2})^3\right] - \frac{2(1+l)}{\pi}(\arcsin p' + p'\sqrt{1-p'^2})\right\}, \tag{8.20}$$

2) for $p' \geq 1$

$$I(\theta) = \frac{a\cos\theta}{2+\alpha l}\left(1 + \frac{4l}{3\pi p'}\right). \tag{8.21}$$

The formula (8.19) already shows that the effect of the condensation coefficient on the intensity of the molecular flow is the same in all directions. Equations (8.20) and (8.21) show still more distinctly that the condensation coefficient changes only the scale of the directivity diagram without changing its shape; this is natural, since it is assumed that $\alpha \neq \alpha(\theta)$ and $\alpha \neq \alpha(r)$.

If the condensation coefficient depends on direction, there will be changes in the shape of $\Phi(z)$ and in the directivity diagram $I(\theta)$. The ratio t of the linear dimensions of the directivity diagram of a molecular beam from a cylinder with a bottom (where the condensation coefficient is α) to the linear dimensions of the Clausing diagram is given by the formula

$$t = \frac{2 + l}{2 + \alpha l} . \tag{8.22}$$

SECTION 9

AN INVESTIGATION OF EVAPORATION IN THE KNUDSEN PRESSURE RANGE

Here we shall consider the geometrical aspect of the problem of determining the saturated vapor pressure by the effusion-weight loss method [Knudsen, 1909 (2,3), 1911 (10)].*

This method is based on a determination of the effusion rates of molecules of the substance in question from a vessel called the effusion chamber. Knudsen suggested using as an effusion chamber a complete analog of the absolutely black body—an isothermal envelope filled with both the substance in question and its vapor and having an effusion aperture in the wall. A method opposite to this is that of Langmuir [1913 (2)] in which the vapor pressure was determined from the rate of evaporation a from an open surface:

$$a = \alpha u (n_s - n)/4, \qquad (9.1)$$

where n_s is the equilibrium concentration of the vapor in particles per cubic centimeter. Equation (9.1) is called the Hertz-Knudsen formula (quoted by Knacke and Stranski, 1956).

According to Langmuir, when one knows the rate of evaporation, one can calculate the vapor pressure p from the formula

$$p = 17.14 \frac{a_m}{\alpha} \sqrt{\frac{T}{M}}. \qquad (9.2)$$

The saturated vapor pressure p is connected with the effusion rate from an ideal Knudsen cell** by the formula (which is a definition of the ideal cell)

$$p = 17.14 a_{me} \sqrt{\frac{T}{M}}, \qquad (9.3)$$

where p is in mm Hg, and a_{me} is the effusion rate in grams per second per centimeter of the area of the effusion aperture.

Melville (1936) and then Marshall, Domte, and Norton (1937) raised the question of the influence of the geometrical surface properties of the evaporating material on the rate of evaporation (this question had already arisen in Knudsen's work). The actual surfaces of bodies are by no means plane. In addition to roughness caused by mechanical, thermochemical, and other kinds of treatment, the processes of evaporation and condensation produce on a surface definite microrelief structures which depend on the condition under which evaporation occurs.†

*A detailed review of methods for the measurement of vapor pressures is given in the monograph by A. N. Nesmeyanov, 1961.

**We note that we understand the ideal Knudsen cell to be the same sort of abstract concept as the ideal black body in optics, so that actual cells are only more or less close approximations to this (see the following portion of the text).

†For these and other facts relating to the behavior of surfaces of single crystals and polycrystalline substances, see the collection of articles edited by G. G. Lemmlein and A. A. Chernov (1959); B. Honigmann (1961); and the collection edited by N. N. Sheftal' (1963).

Specimens for evaporation are frequently taken in the form of powder or chips, for which it is difficult to secure a constant surface state. The shape of the surface can undergo changes not only from experiment to experiment, but also during the course of a single experiment. Therefore it is readily understandable that the effective evaporation coefficient q can also vary over a wide range. This connection between the geometry of the surface and the effective evaporation coefficient follows from the fact that when the surface area is increased there can be an increase in the number of collisions which vapor particles make with it.

As early as 1924 V. A. Fok showed that the illuminance from a surface of arbitrary shape depends only on the outline boundary of the luminous surface, and not on the shape of the surface itself. This theorem holds under the condition that the surface radiates according to Lambert's law and that its reflection coefficient is equal to zero.

Marshall and his collaborators (1937) found an empirical criterion: $\alpha \neq 1$ if the rate of evaporation of a specimen with a smooth surface is less than that of the same specimen when holes have been bored in its surface.

Accordingly, here again we can establish a parallel between photometry and the theory of evaporation, since Fok's theorem and the criterion given by Marshall and his collaborators are really the same thing stated in the languages of two different branches of science. It would be possible to derive the Marshall criterion by using precisely the mathematical apparatus of Fok's work.

Whereas for $\alpha = 1$ there can be no difficulty in principle in determining the vapor pressure from the rate of evaporation, for $\alpha \neq 1$ the situation is more difficult. In the Langmuir method the difficulty arises because α appears in the formula used, Eq. (9.2). In the Knudsen method, since actual effusion cells have Clausing coefficients not equal to unity (see Sections 6,7), the result is that Eq. (9.3) is no longer valid.

A number of methods for determining evaporation coefficients have been proposed which are based on various sorts of comparisons between the results of experiments on evaporation from an open surface (as in Langmuir's method) and experiments on evaporation from a Knudsen cell. Often, however, there are differences of several orders of magnitude between the data of different authors.*

The necessity of evaluating the influence of the geometry of the effusion vessel (in particular the Knudsen cell) has caused the appearance of quite a number of papers aimed at elucidating this influence. We shall consider the main papers of this sort.

In a paper by Stern and Gregory (1957) a formula for the process of evaporation in a Knudsen cell is written as follows:

$$p = p_T + \frac{p_0 S_0}{\alpha S_T}. \tag{9.4}$$

Using the Clausing relation, we can write

$$p_T = p_0\left[1 + \frac{S_0}{S_T}\left(\frac{1}{W} - 2\right)\right], \tag{9.5}$$

where W is the Clausing factor. For $S_0/S_T \ll 1$ we can assume that $p_0 = p_T$. Then Eq. (9.4) becomes the relation obtained by Motzfeld (1955):

$$p_T = p - \left(\frac{1}{\alpha} + \frac{1}{W} - 2\right)\frac{S_0 p_T}{S_T} W_0, \tag{9.6}$$

*We may consider, for example, the data on the evaporation coefficient of cadmium: according to the data from work by Frauenfielder (1950) we have $\alpha = 10^{-3}$; according to Devienne (1953) $\alpha = 0.04-0.6$; according to Rapp and others (1961) $\alpha = 1$; and so on.

where W_0 is the Clausing factor for the aperture of the Knudsen cell.

If $1/W - 2$ is much smaller than $1/\alpha$, and $W_0 = 1$, then Eq. (9.6) takes the simpler form of the equation derived in a paper by Speiser and Johnston (1950):

$$p_T = p - \frac{S_0 p_T}{S_T \alpha}. \tag{9.7}$$

This equation is very widely used, but is not always justified. Even following the logic of Stern and Gregory and of Motzfeld, it is more correct to write

$$p_T = p - \frac{S_0 p_T \cdot W_0}{S_T \alpha}. \tag{9.8}$$

From Eq. (9.8) it is easy to derive a formula for the determination of α from two (or more) measurements of the effective vapor pressure $p_{Ti} = p_{oi}$ $(i = 1, 2 \ldots)$, at various values of W_{oi}, S_{oi}:

$$\alpha = \frac{(p_{T2} W_{02} \cdot S_{02} - p_{T1} W_{01} S_{01})}{S_T (p_{T1} - p_{T2})}. \tag{9.9}$$

As was shown by Lozgachev (1961) the value of α in Eq. (9.9) may be only an effective evaporation coefficient (in Lozgachev's notation, q). Moreover, to increase the accuracy, q must be found under definite conditions of evaporation.

Whitman (1952, 1953) gave the following formula for calculating the vapor pressure from the rate of effusion from a Knudsen cell, $\dot{a}_e$:

$$p = \frac{\dot{a}_e}{\alpha W'} \sqrt{\frac{2\pi RT}{M}} (1 - b'), \tag{9.10}$$

where

$$W' = WW_0/[1 - (1 - \sigma W)(1 - W)],$$

$$b' = (1 - \alpha)[1 - W + W'W(1 - \sigma W_0)/W_0],$$

$$\sigma = \frac{S_0}{S_T},$$

W is the probability of a molecule traveling from the surface of the evaporating substance to the effusion aperture, and W_0, is the probability that a particle arriving at the effusion aperture will pass through it.

For $\sigma \to 0$, $W' = W_0$, we have $b' = (1 - \alpha)$ and Eq. (9.10) takes the form

$$p = \frac{\dot{a}_e}{W_0} \sqrt{\frac{2\pi RT}{M}}. \tag{9.11}$$

As a result of the researches we have listed, and especially of the work of V. I. Lozgachev (1959-1963), who in our opinion has given the most general of the published formulas for the probability of effusion from vessels of arbitrary shape, relations have been found for the dependence of the effective condensation coefficient on the ideal (or "true," in Lozgachev's terminology) and geometrical characteristic of the surface. In this connection it is helpful to consider the results of V. I. Lozgachev, who in his 1961 paper, after deriving the formula

$$p = 17.14 \frac{a'_{me}\sigma}{\alpha\bar{\omega}(\alpha)}\sqrt{\frac{T}{M}} \qquad (9.12)^*$$

writes: "Equation (9.12)** allows us to treat the Knudsen and Langmuir methods on an equal footing, unifying them in a single general method of evaporation into vacuum. We note that this conclusion was reached much earlier by Whitman (1952), but he did not succeed in separating the condensation coefficient from the geometrical factors and arriving at the probability expression (9.16)."†

We shall give the derivation of the formula mentioned for the probability $\bar{\omega}(\alpha)$ of a molecule leaving a vessel of arbitrary shape, and also the derivation of Eq. (9.12). Let (Lozgachev, 1963) a'_{me} be the rate of effusion from a real Knudsen cell, and a_{me} be the rate of effusion for $W_0 = 1$. Then

$$\frac{p_0}{p} = \frac{a'_{me}}{a_{me}W'_0}. \qquad (9.13)$$

Since $a_m = a_{me}\alpha$,

$$a'_{me} = \frac{a_{me}\alpha S_T\bar{\omega}(\alpha)}{S_0}, \qquad (9.14)$$

where $\bar{\omega}(\alpha)$ is the probability of a molecule passing from the surface into the vacuum, averaged over the entire surface of the evaporating substance.

Substituting Eq. (9.14) in Eq. (9.13), we get the formula

$$\frac{p_0}{p} = \frac{\alpha S_T}{W_0 S_0}\bar{\omega}(\alpha), \qquad (9.15)$$

which is equivalent to Eq. (9.12) if we assume that Eq. (9.11) gives not p, but p_0.

We still need to derive the form of $\bar{\omega}(\alpha)$. Let $\bar{\omega}(1)$ be the probability that a particle would leave the effusion vessel if we had $\alpha = 1$. Ordinarily, the process of effusion is limited not by the rate of evaporation, but by τ, the mean free time of a molecule between collisions. The molecule can make collisions either with the wall of the vessel or with the surface of the evaporating substance. Suppose that during this time ν particles evaporate, then the fraction $\bar{\omega}(1)$ of them will escape and not return, and there will be $\nu[1-\bar{\omega}(1)]$ particles remaining in the vapor phase. This brings us to the end of the period τ_1, and a new period starts during which another ν particles will evaporate and $\nu[1-\bar{\omega}(1)]\alpha$ particles will condense.

Accordingly, there will be $\nu[(1-\bar{\omega}(1))(1-\alpha)+1]$ particles in the vapor phase, of which it is possible for only the fraction $\bar{\omega}(1)$ to escape; this means that at the end of the second period τ_2 there will be $\nu[\omega(1)+(1-\alpha)\cdot(1-\bar{\omega}(1))\bar{\omega}(1)]$ particles which have come out of vessel. Continuing this line of reasoning we see that at the end of a period τ_n, when steady state flow has been established, there will be $\nu\{\bar{\omega}(1)+\ldots+(1-\alpha)^{n-1}[1-\bar{\omega}(1)]^{n-1}\cdot\bar{\omega}(1)\}$ particles that have emerged from the effusion vessel. The expression in brackets is the required probability

*Quantities which appear in the formula and have not been defined earlier will be dealt with below.

**In the quotation we have used our notations for formulas and references.

†See below.

$\omega(\alpha)$ and can easily be calculated as the sum of a decreasing geometric progression:

$$\bar{\omega}(\alpha) = \frac{\bar{\omega}(1)}{\alpha + \bar{\omega}(1)(1-\alpha)}. \tag{9.16}$$

Thus the problem of finding $\bar{\omega}(\alpha)$ has been solved, but now we must find $\bar{\omega}(1)$ (see footnote on page 45).

We shall note some aspects of the derivation of Eq. (9.16) which were taken for granted earlier and not mentioned. In order for Eq. (9.16) to be true, even for a monatomic vapor, the entire inside surface of the vessel must be coated (lined) with the substance to be investigated, and in addition it is necessary that the molecular flux in the cell be isotropic. Otherwise, by the beginning of the second period τ_2 there will not be $\nu[1+(1-\bar{\omega}(1))(1-\alpha)]$ particles in the vapor phase, but somewhat more or less depending on the ratio of the areas and the relative positions of the portions of the evaporating surface and of the rest of the inside surface of the vessel, and also depending on the value of the reflection coefficient $(1-\alpha)$ of the remaining part of the inside surface of the cylinder.

For the effective evaporation coefficient q, Lozgachev* gives the formula:

$$q(\alpha) = \alpha + (1-\alpha)\left\{1 - \omega_p(\alpha)\left[\frac{S_{pr}}{S_k} + \frac{S_{dep}}{S_k}\omega_m(q_p)\right]\right\}, \tag{9.17}$$

where $\omega_p(\alpha)$ is the probability for emergence of the molecule from a depression if the surface of the depression has natural roughness; S_{pr} is the area of the prominences, S_{dep} is the apparent area of the depression, S_k is the area of the apparent surface (the projection of the true surface on a given plane), and ω_m is the probability of emergence of a molecule from a depression if the surface of the depression is not rough.

For the special case of a surface covered with grooves of triangular cross section with the angle 2θ at the vertex of the grooves, the value found is

$$\omega_p(\alpha) = \frac{\sin\theta}{[\alpha + \sin\theta(1-\alpha)]}. \tag{9.18}$$

If the evaporator is diffuse,

$$q_p(\alpha) = \frac{\omega_p(\alpha)\,\alpha}{\omega_p(1)}. \tag{9.19}$$

For evaporating surfaces of cubic single crystals or polycrystalline specimens located at an angle φ with the principal faces of the crystal,

$$\omega_p(1) = 0.794.$$

If we assume that S_T, the true surface area of the evaporator, is given by the formula

$$S_T = \frac{S_{dep}}{\sin 45°}$$

*In the sense of Lozgachev's argument, in Eqs. (9.12), (9.14), (9.15), and in the entire text up to the derivation of Eq. (9.16) it would be more correct to write q instead of α. On the relation between q and α see below.

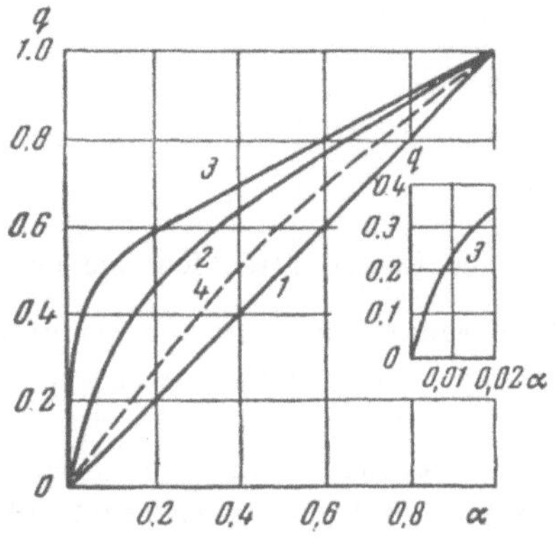

Fig. 20. Effective condensation coefficient q as a function of the ideal condensation coefficient α and the ratio of the apparent area S_k to the true area S_T of the evaporation surface (Lozgachev) 1) $S_T = S_k$; 2) $S_T = 5S_k$; 3) $S_T = 50\ S_k$: 4) $S_{pr} = 0$.

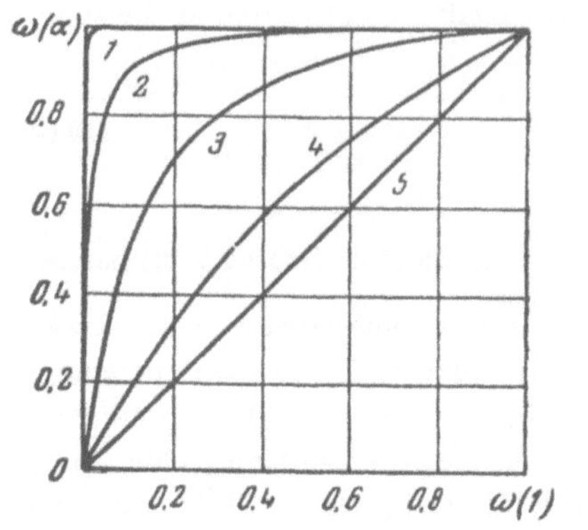

Fig. 21. Dependence of the true probability $\overline{\omega}(\alpha)$ of molecules escaping from the evaporating surface on the hypothetical probability $\omega(1)$, for various evaporation coefficients (Lozgachev) 1) $\alpha = 10^{-4}$; 2) 10^{-2}; 3) 10^{-1}; 4) $5 \cdot 10^{-1}$; 5) 1.

and that in addition

$$S_{pr} = S_{dep} = \frac{S_k}{2},$$

then we can construct the dependence $q = q(\alpha\ S_T/S_k)$ shown in Fig. 21. It is easily seen how greatly q may vary, and consequently how great the error can be in measurements of the vapor pressure when there are relatively small changes of S_T/S_k, if α is sufficiently small. In this same paper Lozgachev suggests several methods for determining the coefficients q and α, on of which has already been pointed out earlier [see Eq. (9.9)].

Figure 21 shows the effect of the evaporation coefficient on the value of $\overline{\omega}(\alpha)$ for various values of $\omega(1)$. Comparing these data with Eqs. (9.15) or (9.12), we can easily estimate the possible experimental error in the determination of the saturated vapor pressure. The actual discrepancy between the data of various experimenters amounts to several hundred percent.

In his 1962 papers Lozgachev derived integral equations for the problem of the passage of molecules through vessels of arbitrary shape. For special cases, in particular for cylinders with membranes, the solutions of the integral equations were also obtained.

TABLE 9.1. Transmission Probability $\vec{W}'$ for a Cylindrical Volume (R = 1; L = 2) with a Small Axial Ideal Effusion Aperture of Radius r (Balson—B; Whitman and Motzfeld—WM)

r	0.1	0.2	0.3	0.4	0.5	1.0
$\vec{W}'$ (B)	0.99530	0.97040	0.93480	0.88906	0.83546	0.51348
$\vec{W}'$ (WM)	0.9906	0.9635	0.9215	0.8685	0.8083	—

In concluding the present section, we present the formula for finding the permeability of a cylinder with a membrane:

$$\vec{W} = \frac{W\dfrac{S_2}{S_1}}{\dfrac{S_2}{S_1} + W\left(1 - \dfrac{S_2}{S_1}\right)}, \tag{9.20}$$

where W is the Clausing coefficient of a tube whose cross section is S_1, and S_2 is the aperture area of a membrane coaxial with the tube. The probability of passing through is calculated according to Eq. (9.20) for a gas moving from the open end of the tube to the membrane. Some of Lozgachev's results (1962) agree with the data of Balson (1961).

We give the result of a calculation by Balson (1961) of the probability $\vec{W}'$ for passage of a molecule along an entire cylinder which has at one end a membrane of radius r and is open at the other end (Table 9.1).

To explain the difference between $\vec{W}$ of Eq. (9.20) and $\vec{W}'$, we give the calculation of the effusion rate of a gas from a cylinder with a membrane.

Suppose that the quantity a mole cm^{-2} sec^{-1} passes through the open end of the cylinder; then, according to Balson, $\pi r^2 a\vec{W}'$ mole/sec passes through the diaphragm, and, according to Lozgachev, the amount passing through the diaphragm is $\pi R^2 a\vec{W}$ mole/sec.

Let us compare the results of calculating the effusion rate a_e' according to Balson and according to Lozgachev for a cylinder L = 2, R = 1, for various values of r, if the rate of arrival of particles is $a = 1000/\pi$ mole cm^{-2} sec^{-1} (Table 9.2).

It is clear from Table 9.2 that we must have the approximate relation

$$\vec{W}'r^2 = \vec{W}R^2,$$

or

$$\vec{W}' = \vec{W}\frac{S_1}{S_2}.$$

When it is necessary to determine the rate of effusion from a Knudsen cell containing material with an evaporation coefficient α, this can be done either according to the formulas of Section 7 or according to the formulas given in the present section (the result of Lozgachev or of Whitman and Motzfeld), or else, finally, according to Balson. In Table 9.3 we give the variation of the coefficient $\vec{W}'_{(\alpha)}$ as a function of α. Following Balson (1961) we give this dependence in the form of a table of $\vec{W}'(\alpha)/\vec{W}'(1)$; for comparison we also give the the results of calculations according to the Motzfeld formula. Balson points out that calculation according to Whitman and Motzfeld gives practically identical results.

TABLE 9.2. Comparison of the Results of Calculations of the Effusion Rate a_e' According to Lozgachev (L) and According to Balson (B)

r	0.1	0.2	0.3	0.4	0.5	1.0
a_e' (B)	10	39	84	142	208	513
a_e' (L)	10	38	83	143	203	514

Until very recently it has been held by researchers working on the evaporation of comparatively nonvolatile substances (i.e., substances which must be heated to high temperatures in order for evaporation effects to be noted) that it is necessary in effusion experiments to

TABLE 9.3. Effect of Evaporation Coefficient α on the Rate of Effusion from a Cylindrical Container with a Small Ideal Effusion Aperture. Length of Container L = 2; Radius R = 1 (M–according to Motzfeld; B–according to Balson)

α	$\vec{W}'(\alpha)/\vec{W}'(1)$ $(r/R = 0.1)$		$\vec{W}'(\alpha)/\vec{W}'(1)$ $(r/R = 0.2)$	
	B	M	B	M
0.9	0.9996	0.9990	0.9948	0.9960
0.8	0.9992	0.9975	0.9884	0.9905
0.7	0.9986	0.9955	0.9803	0.9840
0.6	0.9979	0.9933	0.9697	0.9750
0.5	0.9968	0.9900	0.9552	0.9625
0.4	0.9952	0.9850	0.9342	0.9460
0.3	0.9926	0.9770	0.9013	0.9180
0.2	0.9874	0.9615	0.8420	0.8660
0.1	0.9720	0.9095	0.7031	0.7425
0.01	0.7463	0.5050	0.1767	0.2076

have a Knudsen cell which behaves indifferently with respect to the vapor. This belief takes for granted that it is possible for such indifferent partners (substance to be investigated and substance of which the Knudsen cell is made) to exist. There is, however, a steadily increasing amount of experimental information which convinces us that there are no such pairs of materials. Also, on the other hand, there are no proofs that such a relation of indifference is possible. It would evidently be more correct to speak of the degree and nature of the physical and chemical interaction between the evaporated substance, the vapor, and the inside surface of the Knudsen cell. The formulas of the present section do not take into account the possibility of this sort of interaction.

It must also be pointed out that all of the material presented here assumes that the vapor investigated is monatomic, or at any rate, that the molecules have a unique structure which does not change in collision with the walls of the chamber.

SECTION 10

EQUATIONS FOR THE PROCESS OF ESTABLISHING EQUILIBRIUM IN A KNUDSEN CELL

We shall derive the equation for the process of vapor saturation in a Knudsen cell when among the particles of the vapor phase there are not only monomers but also particles of more complex composition (Lyubitov, 1963; Jepsen and Somorjai, 1963).

The analogous problem for a vapor of purely monomeric composition has been solved by L. P. Firsova (1962).

The case investigated is that in which one can neglect the number of collisions between particles in comparison with the number of collisions that the particles make with the walls of the effusion chamber. Consequently, there can be the following causes for the appearance of particles of type i in the vapor phase:

1) evaporation at the rate a_i; 2) processes of dissociation when particles of type j make collisions with the walls (we denote the dissociation rate constant by k_{ij}).

Vanishing of particles i from the vapor phase is due to (1) their condensation and (2) dissociation at the wall (we denote the particle disintegration rate constant due to these processes by k_{ii}).

It is possible to connect the quantities k_{ii} and k_{jj} by means of the stoichiometric dissociation coefficients at the walls.

To obtain a simpler and more precise statement of the problem we assume that the Knudsen cell is spherical and that its inner surface is covered with the one-component substance to be investigated. This condition assures the necessary isotropic character of the molecular flows inside the cell.

This problem is described by the following system of differential equations:

$$\frac{dn_i}{dt} = a_i + k_{ij} n_j, \tag{10.1}$$

where t is the time and n_i is the concentration of particles of type i in the vapor.

We solve this system by the operator method (cf. Mikusinski, 1959). Replacing d/dt by D, we get a system of inhomogeneous first order linear equations:

$$a_i = Dn_i - k_{ij}n_j. \tag{10.2}$$

If $i > j$, then $k_{ij} = 0$, which makes the solution of the system (10.2) much easier.

As an example let us consider the case in which there are three types of particles in the vapor: A_1, A_2, A_3; their respective rates of evaporation are a_1, a_2, a_3. We assume that the following reactions occur (we indicate the transition of particles into the condensed phase by parentheses):

$$(A_1) \rightleftarrows A_1, \tag{10.3}$$

$$(A_2) \rightleftarrows A_2, \tag{10.4}$$

$$A_2 \rightarrow (A_1) + A_1, \qquad (10.5)$$

$$A_2 \rightarrow 2\,(A_1), \qquad (10.6)$$

$$(A_3) \rightleftarrows A_3, \qquad (10.7)$$

$$A_3 \rightarrow (A_1) + A_2, \qquad (10.8)$$

$$A_3 \rightarrow (A_2) + A_1. \qquad (10.9)$$

The system of equations (10.2) is now written in the form:

$$\begin{aligned} a_1 &= (D + k_{11})\, n_1 - k_{12} n_2 - k_{13} n_3, \\ a_2 &= (D + k_{22})\, n_2 - k_{23} n_3, \\ a_3 &= (D + k_{33})\, n_3. \end{aligned} \qquad (10.10)$$

The coefficients k_{ij} for the reactions shown in Eqs. (10.3)–(10.9) are of the form

$$k_{11} = (S_o + \alpha S)\sqrt{RT}/\sqrt{2\pi M}, \qquad (10.11)$$

$$k_{22} = (S_o + \delta S)\sqrt{RT}/\sqrt{4\pi M}, \qquad (10.12)$$

$$k_{33} = (S_o + \tau S)\sqrt{RT}/\sqrt{6\pi M}, \qquad (10.13)$$

$$k_{12} = (\delta' + 2\delta'')\, S\sqrt{RT}/\sqrt{4\pi M}, \qquad (10.14)$$

$$k_{13} = S\tau''\sqrt{RT}/\sqrt{6\pi M}, \qquad (10.15)$$

$$k_{23} = S\tau'\sqrt{RT}/\sqrt{6\pi M}, \qquad (10.16)$$

where S_0 is the area of the effusion aperture, S is the area of the inside surface of the effusion cell, δ is the coefficient of condensation in the reaction (10.4), δ' is the coefficient of dissociative condensation by the reaction (10.5), δ'' is the coefficient of dissociative destruction of the dimer according to the reaction (10.6), τ is the coefficient of condensation of the trimer according to the reaction (10.7), τ' is the coefficient of dissociative condensation of the trimer according to the reaction (10.8), and τ'' is the coefficient of dissociative condensation according to the reaction (10.9).

Solving the system (10.10) and using the formulas of operational calculus to return from the transforms to the original functions, we get:

$$\begin{aligned} n_1 = a_1 [1 - \exp(-k_{11}t)]/k_{11} &+ a_2 k_{12} [1/k_{11}k_{22} - \exp(-k_{11}t)/k_{11}(k_{22} - k_{11}) \\ &- \exp(-k_{22}t)/k_{22}(k_{11} - k_{22})] + a_3 k_{12} k_{23} [1/k_{11}k_{22}k_{33} \\ &- \exp(-k_{11}t)/k_{11}(k_{22} - k_{11})(k_{33} - k_{11}) \\ &- \exp(-k_{33}t)/k_{33}(k_{11} - k_{33})(k_{22} - k_{33})], \end{aligned} \qquad (10.17)$$

$$n_2 = a_2[1 - \exp(-k_{22}t)]/k_{22} + a_3k_{23}[1/k_{22}k_{33} - \exp(-k_{22}t)/k_{22}(k_{33}-k_{22}) - \exp(-k_{33}t)/k_{33}k_{22}-k_{33})], \tag{10.18}$$

$$n_3 = a_3[1 - \exp(-k_{33}t)/k_{33}. \tag{10.19}$$

For $t \to \infty$ Eqs. (10.17)–(10.19) take the form

$$n_1(\infty) = a_1/k_{11} + a_2k_{12}/k_{11}k_{22} + a_3k_{12}k_{23}/k_{11}k_{22}k_{33}, \tag{10.20}$$

$$n_2(\infty) = a_2/k_{22} + a_3k_{23}/k_{22}k_{33}, \tag{10.21}$$

$$n_3(\infty) = a_3/k_{33}. \tag{10.22}$$

Equations (10.17)–(10.22) are sufficient for calculation of the concentrations n_i. One could increase the number of equations by planning additional effusion experiments using cells of this type with various values of S_0. In this connection one must keep in mind certain restrictions on the correctness of the Knudsen method. It is easy to relate the quantities n_i to the pressure which is established in the cell. It is not possible, however, for us to measure this pressure directly. We can measure the total and partial effusion rates, and then use more or less justified formulas to calculate the corresponding pressures – for example, by using the working formula set up by Motzfeld (1955) which is valid for a vapor consisting of particles of a single type [cf. Eq. (9.6)]:

$$p_T = p - (1/\alpha + 1/W - 2)\, S_0 p_T\, W_0/S_T\ . \tag{10.23}$$

It may be noted, first, that from the expressions (10.17)–(10.22) one can easily derive the formulas given by L. P. Firsova (1962), and second, that Eqs. (10.17)–(10.22) can be easily converted into expressions for a vapor consisting of particles of any number of types.

If there are different processes going on in a particular system, they can be taken into account by the method indicated. It is not difficult to transform equations of the type (10.23) so as to obtain the relationship for our variables n_i. The Clausing coefficients known up to very recent times have not taken into account the chemical aspects of the interaction of the vapor phase with the walls of the effusion vessel. Moreover, Eq. (10.23) is rather complicated to use, and therefore in practical research one can either make some assumptions and simplifications, or carry out studies of evaporation by means of vessels in which the effect of the geometry can be more easily calculated. There have been particularly thorough studies of cylindrical effusion vessels.

Jepsen and Somorjai (1963) analyzed the kinetics of processes occurring in a Knudsen cell. Their analysis was very similar to the one presented above. The principal distinctive features of their work are the following: 1) The model is complicated by the assumption of an intermediate adsorbent layer; 2) it is assumed that reaction surfaces represent the limiting stage of the evaporation process; 3) the problem of a series or parallel reaction mechanism on the evaporating surface is distinctly formulated and solved.

We will give a brief outline of their results, with minor changes in notation.

Let β be the rate of diminution of the number of particles in the enclosure due to effusion, whereupon in the approximation of Whitman (1953) and Motzfeld (1955)

$$\beta_i = \frac{W' S_0}{1 - W'\sigma}\sqrt{\frac{RT}{2\pi M_i}}$$

where σ and W' have the same meaning as in Eq. (9.10).

For $S_0 \to 0$ we have also $\beta_i \to 0$. As S_0 is increased, $W'\sigma \to 1$ and $\beta_i \to \infty$. Consequently, β_i varies from 0 to ∞.

Following the work of Jepsen and Somorjai, the analysis is carried out for a one-component substance capable of yielding monomers and dimers in the vapor phase.

a) Parallel reaction mechanism on the surface. In this case the following reactions, for example, are possible:

$$(A) \underset{k_2}{\overset{k_1'}{\rightleftarrows}} A_T \tag{10.25}$$

$$(A_2) \underset{k_4}{\overset{k_3'}{\rightleftarrows}} A_{2T} \tag{10.26}$$

$$A_T \underset{k_6}{\overset{k_5}{\rightleftarrows}} A \tag{10.27}$$

$$A_{2T} \underset{k_8}{\overset{k_7}{\rightleftarrows}} A_2, \tag{10.28}$$

where A_T and A_{2T} are the surface compounds of A and A_2 and the remaining notation is the same as before.

For steady state processes we can write

$$\frac{dn_T}{dt} = k_1' n_{(A)} - k_2 n_T - k_5 n_T + k_6 n_1 = 0 \tag{10.29}$$

$$\frac{dn_{(A)}}{dt} = k_1' n_{(A)} + k_2 n_T = 0 \tag{10.30}$$

$$\frac{dn_{2T}}{dt} = k_3' n_{(A_2)} - k_4 n_{2T} - k_7 n_{2T} + k_8 n_2, \tag{10.31}$$

where n_T and n_{2T} are the concentrations of the particles A_T and A_{2T} in the adsorbed layer, n_1 and n_2 are the concentrations of particles A and A_2 in the gaseous phase, $n_{(A)}$ and $n_{(A_2)}$ are the constant concentrations of the particles A and A_2 in the condensed state.

Hence, with $k_1 = k_1' n_{(A)}$ and $k_3 = k_3'{}_{(A_2)}$, we obtain

$$\frac{1}{n_1} - \frac{1}{n_{01}} = \beta_1 \frac{k_2 + k_5}{k_1 k_5} \tag{10.32}$$

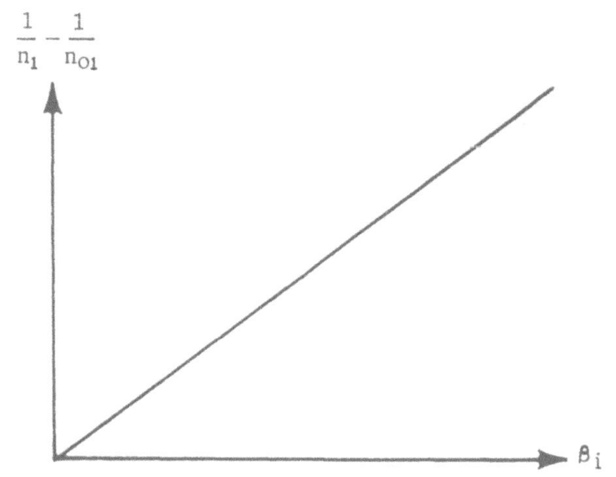

Fig. 22. Variation in the difference of the reciprocal concentrations with size of the effusion aperture in a Knudsen cell, assuming a parallel mechanism of monomer and dimer evaporation (Jepsen and Somorjai).

$$\frac{1}{n_2} - \frac{1}{n_{02}} = \beta_2 \frac{k_4 + k_7}{k_3 k_7} \tag{10.33}$$

where n_{01} and n_{02} are the equilibrium concentrations (for $\beta = 0$). The behavior of the dependence of $1/n_i - 1/n_{0i}$ on β_i is shown in Fig. 23 ($i = 1$ corresponds to monomers, $i = 2$ to dimers).

The slope corresponds to the coefficient β_i in Eqs. (10.32) and (10.33).

Consequently, a parallel mechanism for the evaporation of monomers and dimers is characterized by a linear dependence of $1/n_i - 1/n_{0i}$ on β_i.

b) Evaporation accompanied by a series, or chain, reaction on the surface:

$$(A) \underset{k_2}{\overset{k_1'}{\rightleftarrows}} A_T \tag{10.34}$$

$$(A) + A \underset{k_4}{\overset{k_3}{\rightleftarrows}} A_{2T} \tag{10.35}$$

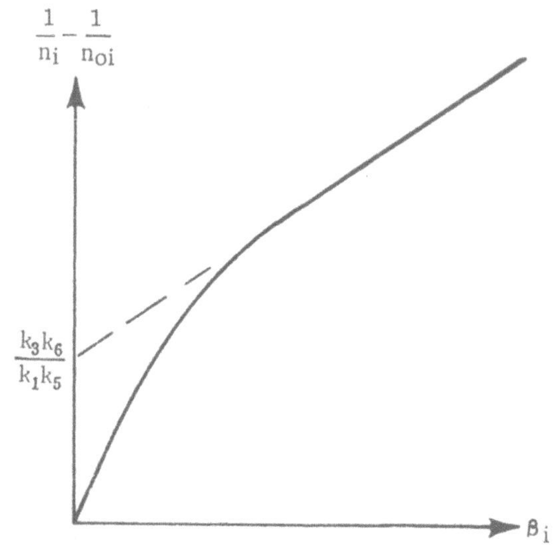

Fig. 23. Difference in the reciprocal concentrations of monomers (under steady state and equilibrium conditions) as a function of the size of an effusion aperture in a Knudsen cell, assuming a chain mechanism of surface reaction (Jepsen and Somorjai).

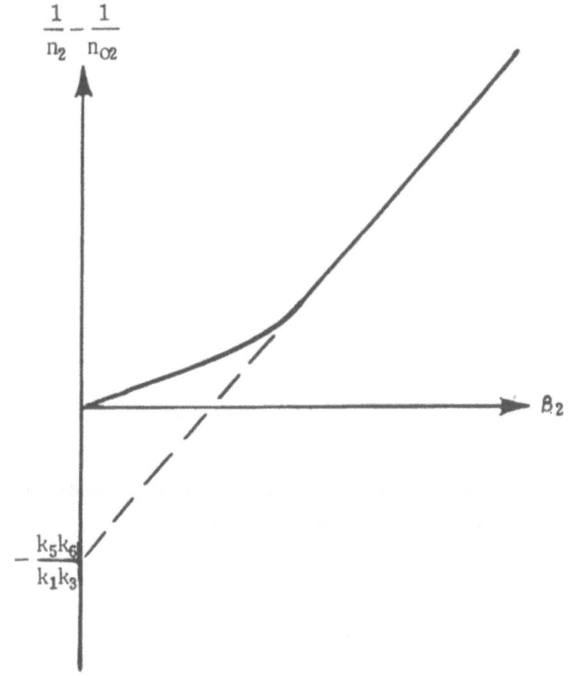

Fig. 24. Difference in the reciprocal concentrations of dimers (under steady state and equilibrium conditions) as a function of the size of an effusion aperture in a Knudsen cell, assuming a chain mechanism of surface reaction (Jepsen and Somorjai).

$$A_T \underset{k_6}{\overset{k_5}{\rightleftarrows}} A \tag{10.36}$$

$$A_{2T} \underset{k_8}{\overset{k_7}{\rightleftarrows}} A_2 , \tag{10.37}$$

where the "chain link" is the reaction (10.35), all the remaining reactions being as before. It would not be too troublesome to write the system of kinetic equations for these reactions, whereupon the equations for A and A_2 would be obtained in the following form under steady state conditions:

$$n_1 = \frac{k_1 k_5 [k_8 - (k_4 + k_7)\theta]}{k_5 k_6 [(k_4 + k_7)\theta - k_8] + (\beta + k_6)[k_3 k_4 \theta - (k_4 + k_7)\theta K + k_8 K]} \tag{10.38}$$

$$n_2 = \frac{-k_1k_3(\beta + k_6)}{k_5k_6\,[(k_4 + k_7)\,\theta - k_8] + (\beta + k_6)\,[k_3k_4\theta - (k_4 + k_7)\,\theta K + k_8K]}, \qquad (10.39)$$

where

$$\theta = \frac{\beta + k_8}{k_7},$$

$$K = k_2 + k_3 + k_5,$$

and it is assumed that $\beta_1 = \beta_2 = \beta$.

The equilibrium concentration is obtained when $\beta = 0$:

$$n_{01} = \frac{k_1k_5}{k_2k_6}$$

$$n_{02} = \frac{k_1k_3k_7}{k_2k_4k_8}.$$

For sufficiently large β, Eqs. (10.38)–(10.39) give

$$\frac{1}{n_1} - \frac{1}{n_{01}} \cong \frac{1}{k_1k_5}\left(\frac{k_3k_6k_7}{k_4 + k_7} - \frac{k_3k_4k_7k_8}{(k_4 + k_7)^2}\right) + \left(K - \frac{k_3k_4}{k_4 + k_7}\right)\beta \qquad (10.40)$$

and

$$\frac{1}{n_2} - \frac{1}{n_{02}} \cong -\frac{k_5k_6}{k_1k_3} + \frac{k_4k_5(k_8 - k_6)}{k_1k_3k_7} + \left[\frac{K}{k_1k_3}\left(1 + \frac{k_4}{k_7}\right) - \frac{k_4}{k_1k_7}\right]\beta. \qquad (10.41)$$

Graphs of the dependence described by Eqs. (10.40) and (10.41) are shown in Figs. 23 and 24, respectively.

These dependences could be observed and recorded, for example, with a mass spectrometer. Their form could clearly be used as a measure of the mechanism of the reactions accompanying evaporation. The valid assumption is made in Figs. 23 and 24 that

$$k_5,\ k_6,\ k_7,\ k_8 \gg k_1,\ k_2,\ k_3,\ k_4.$$

The actual evaporation mechanisms may prove to be far more complex than the simple models proposed above. However, other models can be investigated on the basis of the notions developed herein. It is very likely that the behavior of many reactions will be sensitive to variations in the geometrical characteristics of the cell (size of the effusion aperture, shape, dimensions, and position of the evaporating material, etc.), cell temperature (which might possibly vary over different parts of the surface), and cell composition.

SECTION 11

CHEMICAL REACTIONS IN LONG CAPILLARIES*

In ascertaining the limits of applicability of the concepts used herein, we have noted (cf. Sec. 3) that molecular flows are established at relatively large pressures in narrow capillaries, as well as in vessels of rather large size at high degrees of rarefaction. There are different mathematical ways of solving problems for capillaries of different lengths, both in the case of purely physical interaction between molecular flow and the walls, and in the case of chemical interaction.

The legitimacy of the assumption of infinite length is illustrated by some parameters for one gram of activated charcoal (cf. Wheeler, 1951):

area of internal surface S_g	$500-1500$ m^2,
volume of pores V_g	$0.6-0.8$ cm^3,
average radius of pores r	$10-20$ A
total length of pores L	10^7 km.

Consequently, in this case the dimensionless length of the pores is 10^{19}!

In the interest of simplicity, for the area S_g of the pore surfaces in one gram of catalyst we assume the value determined experimentally by one of the methods known at the present time (cf. Emmett, 1948).

Let the area of the external geometrical macroscopic surface of a grain of catalyst be S_x, and let the number of pores per unit external surface be n_p; the total volume of the grain is V_p; d_p is the density of the grain.

Theoretically the pore volume is given by

$$S_x n_p \pi r^2 L. \tag{11.1}$$

Let the porosity ψ be the fraction of the volume of an individual grain which consists of the volume of the pores. Then the experimentally determined volume of the pores in the entire grain of catalyst will be

$$V_p \cdot \psi = V_p d_p V_g. \tag{11.2}$$

Equating the expression (11.1) to the right-hand side of (11.2), we get

$$S_x n_p \pi r^2 L = V_p d_p V_g. \tag{11.3}$$

The surfaces of the catalyst can have a certain degree of roughness P (a number which shows by what factor the true surface of the catalyst would be decreased if all of the microscopic roughnesses were smoothed away).

In setting up the equation for the area of the surface it is necessary to remember that pores intersect each other and this decreases the surface area of an individual pore by a factor of approximately $(1-\psi)$. The equation

*The principal literature relating ro this topic includes: Zel'dovich (1939); Thiele (1939); Wheeler (1951, 1955); Chambré (1960); and Tinkler and Metzner (1961).

for the surface area of the pores of the catalyst can be written in the form

$$S_x n_p 2\pi r P (1 - \psi) L = V_p \cdot d_p \cdot S_g. \tag{11.4}$$

From Eqs. (11.3) and (11.4) we find the equation for the average pore radius

$$r = 2 V_g P (1 - \psi)/S_g. \tag{11.5}$$

It can be shown that the equation

$$n_p = \psi/\pi r^2 \sqrt{2} \tag{11.6}$$

holds approximately, and after substituting this in (11.3) we get

$$L = \sqrt{2}\, V_p/S_x. \tag{11.7}$$

Let us consider the principal physical factors which determine the reaction rate in porous catalysts and derive the differential equation for this process.

The treatment of the chemical reaction in the pores is complicated by the fact that different parts of a pore have different degrees of accessibility. Owing to this, the ratio between the processes in the kinetic, transitional, and diffusion regions is altered in comparison with the case of the reaction at a surface, all of whose points are equally accessible to the reagents. Academician Ya. B. Zel'dovich [1939 (1)] writes on this point:

"In this case the accessibility is different for different parts of the active surface which lie at greater or less depths inside a segment of catalyst, and we are in the kinetic region only when the rate of the chemical reaction is smaller than the diffusion rate of the substance to the parts of the surface to which access is most difficult.

"In the diffusion region the rate of transport to the most easily accessible parts of the active surface, which are located on the outside of the segment of catalyst, must be smaller than the rate of reaction at the portions of the surface.

"The transition region is remarkably extended, and takes in the range in which the rate of reaction changes by a factor equal to the ratio (rate of diffusion or other type of transport of the substance to the most easily accessible portions of the active surface): (rate of transport to the parts of the surface to which access is most difficult)."

It is clear that the decisive quantity is the ratio between the reaction and transfer rates of the reagent and the products of reaction. When the reaction rate is large, only a small fraction of the surface takes part in the reaction. If the reaction is slowed down, then particles of the reagent can diffuse into practically the entire depth of a catalyst grain. Let us consider this interconnection in a quantitative way.

Let k be the reaction rate constant, or the number of molecules which react at unit surface area of the catalyst per unit time when the concentration of the reacting molecules at the surface is equal to unity. For a substance i we can write

$$\frac{dn_i}{dt} = k c_i^m dS, \tag{11.8}$$

where m is the reaction order exponent and dn_i/dt is the number of molecules which react per second at the area dS when it is in contact with the reagent at concentration c_i.

Ordinarily the temperature dependence of k is that given by the Arrhenius equation.

In an experiment with a nonporous monolithic catalyst, whose surface will be in contact with a reagent at constant concentration c_i, it is not hard to determine the constant k. To make any further progress it is necessary to derive the equation for the distribution of c_i throughout the interior of a catalyst grain.

If U_i' is the rate of advance of the reacting substance in a given part of the grain and U_i'' is the rate of the reverse flow, then the difference $U_i'-U_i''$ of these quantities must be equal to the reaction rate in the region in question:

$$U_i' - U_i'' = \int kc_i^m dS. \tag{11.9}$$

When the reaction points are close to each other, we can write Eq. (11.9) in the form

$$-dU_i = kc_i^m dS. \tag{11.10}$$

Within the framework of the theory of molecular flow the only cause of material transfer is Knudsen diffusion, and therefore U_i in Eq. (11.10) is given by the formula

$$U_i = -\pi r^2 D dc_i / dx. \tag{11.11}$$

Differentiating (11.11) and remembering that

$$dS = 2\pi r\, dx, \tag{11.12}$$

we get the Fick equation for Knudsen diffusion with allowance made for the effect of chemical reaction:

$$\frac{d^2c_i}{dx^2} = \frac{2k \cdot c_i^m}{rD}. \tag{11.13}$$

A chemical interaction between a molecular flow and the walls of a channel along which it is moving can also be treated in terms of the theory of integral equations. For the concentration distribution $n_i(x)$ of substance i along the channel in the direction of the x axis one can easily derive (in the same way as in Sections 7 and 12) an integral equation

$$n_i(x) = n_{i,0}(x) + n_{i,L}(x) + (1-\alpha)\int_0^L K(|x-x'|)\, n_i(x')\, dx', \tag{11.14}$$

which is valid when there is no inverse reaction.

In Eq. (11.14) we have used the following notation: $n_i(x)$ is the number of molecules of reagent which strike a unit area of the surface at the point x per unit time ($n_i = \text{const}\ c_i$): $n_{i,0}(x)$ is the number of molecules which enter the pore from the side $x = 0$ and make their first collision with the surface between x and $x + dx$, and is measured in the same units as $n_{i,0}$; $n_{i,L}(x)$ is the same quantity as $n_{i,0}$, but for molecules which come into the pore from the other end; α is the probability of the direct reaction; $K(|x-x'|)dx$ is the probability that a molecule which comes from the point X' will arrive directly at a band of width dx at the point x.

Chambré (1960) gives, in addition to the derivation of an integral equation of the type (11.14), the derivation of integral equations for the lengthwise distribution along the channel of reaction products and inert particles when both direct and inverse reactions are occurring. Chambré showed that only in certain cases are these integral equations equivalent to the less general differential equations of the type (7.2) and (11.13).

For example, Eq. (7.2) gives the same result as Eq. (11.14) for any value of α,* but not for all values of L and r – it does so only when we have the inequality

$$0 \leqslant L/2r < 4.$$

The differential equation (11.13) is equivalent to the integral equation (11.14), provided two conditions hold:

$$1)\ \alpha \ll 1 \quad \text{and} \quad 2)\ \frac{\alpha^{1/2}L}{2r} \gg 1.$$

We shall give the result of the solution of the equation (11.13), or, what amounts to the same thing for $L/r = l \to \infty$, that of Eq. (11.14), according to Zel'dovich (1939). The boundary conditions are $c_i = c_{i,0}$ for $x = 0$; $c_i = 0$ for $x \to \infty$.

From dimensional considerations and in view of the fact that the boundary conditions do not contain any quantities with the dimensions of length, we write the general form of the solution:

$$c_i = c_{i,0}\varphi\left(x\sqrt{\frac{k}{Dr}};\ c_{i0,}\right). \tag{11.15}$$

It is clear that the form of φ can be determined only by direct integration of Eq. (11.13). Without doing this, however, we can determine the effective depth of the reaction zone:

$$\varkappa \sim \sqrt{\frac{Dr}{k}}. \tag{11.16}$$

From this we have

$$U_i \sim \frac{k\varkappa}{r} \sim \sqrt{\frac{Dk}{r}}.$$

Whereas according to the law of Arrhenius

$$k \sim e^{-\frac{Q}{RT}}$$

(where Q is the true heat of activation of the reaction), in our case of the transition region

$$U_i \sim \sqrt{k} \sim e^{-\frac{Q}{2RT}}, \tag{11.17}$$

i.e., the observed "heat of activation" can be smaller by a factor of two than the true heat of activation of a heterogeneous chemical reaction.

In a similar way Zel'dovich showed that the order of the reaction in the transition region is an average value between the first order in the diffusion region and the true m-th order in the kinetic region. For example,

*For the process of evaporation α is the condensation coefficient (or the evaporation coefficient).

for a reaction of zeroth order:

$$U_i \sim c_{0,i}^{\frac{1}{2}} \sqrt{DkS} \qquad (11.18)$$

If we alter somewhat the boundary conditions for the solution of Eq. (11.13) to $c_i = c_{0,i}$ for $x = 0$ and $dc_i/dx = 0$ for $x = L/2$, then (according to Wheeler) we can get the desired expression m(m = 0, 1, 2, and so on).

Omitting the not very complicated calculations, we give the solution for a reaction of first order (m = 1) in final form:

$$c_i = c_{0,i} \frac{\coth(h - hx/L)}{\coth(h)}, \qquad (11.19)$$

where h is a dimensionless quantity given by the formula

$$h = L\sqrt{2kc_0^{m-1}/rD}, \qquad (11.20)$$

which for m = 1 gives

$$h = L\sqrt{2k/rD}. \qquad (11.21)$$

Figure 25 shows the dependence of $c_i(x)/c_{i,0}$ on x/L for various values of the parameter h which characerizes the activity of the catalyst.

Figure 26 shows curves for the dependence of the apparent rate constant of the process on the reciprocal of the temperature, for a hypothetical catalyst of various degrees of porosity which has a true activation energy of 20 kcal [cf. the data of Zel'dovich, Eq. (11.7)].

The upper curve corresponds to small h or [according to Eq. (11.21)] large r, the middle curve is for the case h = 0.2 at 425°K, and the bottom curve holds for h ≥ 2.

It is of practical importance to define the accessible part f of the surface, which is found from the formula

$$f = \frac{1}{h} \tanh(h), \qquad (11.22)$$

which gives for h > 2

$$f = 1/h. \qquad (11.23)$$

$$U_i' = \pi r c_{0i} \sqrt{2rkD}, \qquad (11.24)$$

and for h < 1

$$f = 1 - h^2/4. \qquad (11.25)$$

Since f depends only on the dimensionless parameter h, the function f(h) depends only on the relationship between r, k, D, L [cf. Eq. (11.21)].

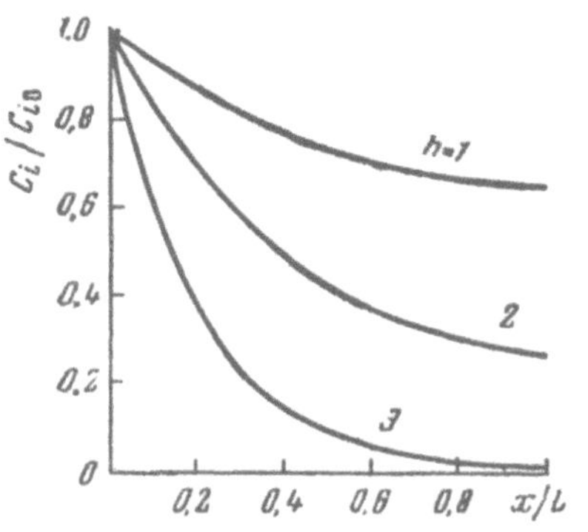

Fig. 25. Decrease in concentration of reacting substance with increasing distance x from the entrance to the pore (Wheeler).

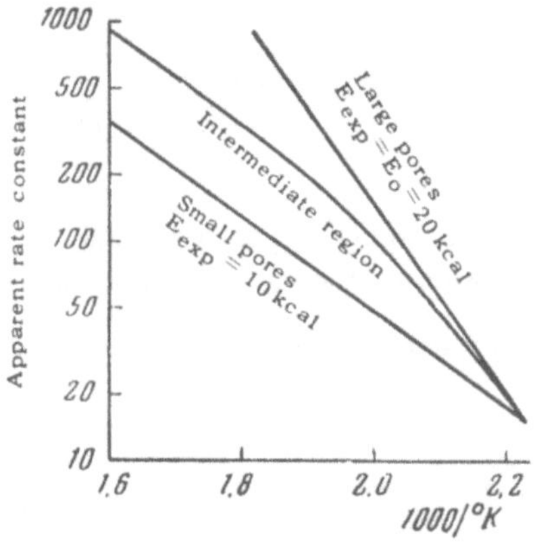

Fig. 26. Influence of the factor h on the apparent heat of activation of a reaction (Wheeler).

We may say that a fast reaction, or a reaction with a catalyst with small pores, implies $h > 2$.

Let us also consider some equations which connect $k(\alpha)$ and h and bring out the physical meaning of h. The observed rate U of the reaction at the surface S will be

$$U = \alpha S u c_i / 4. \tag{11.26}$$

From a comparison of Eq. (11.26) and Eq. (11.8) it is clear that

$$k = \frac{\alpha u}{4}. \tag{11.27}$$

Using Eqs. (3.21), (11.21). (11.27), we get for the quantity h the expression

$$h = \frac{L}{2r}\sqrt{3\alpha}. \tag{11.28}$$

We could deal in a similar way with reactions of other orders ($m = 0$, and so on) and also with a number of questions about the selectivity and contamination of catalysts (Wheeler, 1951).

Unlike Zel'dovich (1939) and Thiele (1939), Wheeler has given methods for the experimental determination of h, and consequently of k, from a separate determination of the activity. We give without derivation Wheeler's result for a static system:

$$h \tanh(h) = \frac{\Phi^2}{18D} \cdot \frac{0.693}{t_{0.5}} \cdot \frac{V_R}{V_g W_g}. \tag{11.29}$$

Here Φ is a characteristic dimension – for example, the diameter of a catalyst grain; $t_{0.5}$ is the time for fifty-percent conversion; V_r is the total volume of the reactor; and W_g is the amount of catalyst in grams.

For the example of cylindrical tubes of infinite length we have seen how the geometrical characteristics of a catalyst can affect the course of a chemical transition.

Actually, of course, catalysts are more complex structures than bunches of intermingled capillaries (cf. Villet and Wilhelm, 1961), and the physical and chemical processes which occur are actually much more complicated than those we have considered. Even the relations given, however, enable us to explain and predict a number of practical results in a comparatively simple way.

We may regard a number of researches of the most recent years as an outgrowth of Knudsen's work of 1910, in which he studied the motion of molecular flows in nonisothermal tubes. A paper by Schilson (1957) gives a calculation of the temperature distribution and the concentration gradient in a grain of a porous catalyst; Preter (1958) observed experimentally the temperature gradient in a grain; and the work of Tinkler and Metzner (1961) gives a theoretical treatment of the chemical reaction rate in a nonisothermal catalyst grain and gives a survey of the available information on this problem.

SECTION 12

CALCULATION OF THE EFFUSION FROM A CYLINDER FOR A VAPOR OF DIMER-MONOMER COMPOSITION

Knudsen developed an effusion method for determining the pressure at saturation of a vapor whose particles are chemically neutral both in relation to the walls of the effusion chamber and in relation to each other.

In a paper by Frazer (1931) it is demonstrated that a Knudsen cell can be used to investigate the reactions of dissociation of the particles of a vapor phase. We note that in this sort of work we are obliged to keep in mind all of the warnings and stipulations which have been made in the foregoing pages, in particular in Sections 9, 10 and 11, with reference to a one-component vapor, which were not known at the time of Frazer's work.

Nevertheless, in the case of an ideal Knudsen cell and equal evaporation coefficients for the components of the vapor phase we are considering, we can use some of the relations which are derived by Frazer (1931).

If p is the total pressure of the gas undergoing dissociation, then the partial pressures p_1 of the monomer particles and p_2 of the dimer particles are given by the formulas:

$$p_1 = 2\varkappa' p/(1+\varkappa'), \tag{12.1}$$

$$p_2 = (1-\varkappa')\, p/(1+\varkappa'), \tag{12.2}$$

where $\varkappa'$ is the degree of dissociation of dimers into monomers.

If from a measurement of the rate of effusion we have calculated the effective vapor pressure p_{cal} on the assumption that there is no dissociation, then to determine $\varkappa'$ we can use the relation

$$2\varkappa'/(1+\varkappa') = p_1/p = 3.41\,(1 - p_{cal}/p). \tag{12.3}$$

Equations (12.1)–(12.3) enable us to calculate the values of the heat of reaction for dissociation and other thermochemical quantities from the data of effusion experiments.

The assumptions with which we idealize our problem are rarely actually justified; the condensation coefficients for the monomer and the dimer are not necessarily the same, and the effusion aperture is a channel of some extent which does not interact in the same way with the various substances effusing from the cell. To include completely the effect of these factors in an actual cell is a complicated problem which has not been entirely solved. These considerations lead us to think about other possibilities for the solution of the given problems. For example, it is possible to use a cylindrical effusion cell, which has a simpler geometry than a Knudsen cell.

Let us consider the evaporation from a cylinder (see Fig. 9) in which the particles evaporating from the bottom and the walls of the cylinder include both monomers and dimers. An example is the case of the vapor of sodium chloride, which consists of NaCl, Na_2Cl_2, and Na_3Cl_3 (see Miller and Kush, 1956; Milne, 1958; and Gorokhov, Khodeev, and Akishin, 1958).

*At the present time, mass spectroscopy is becoming more and more important in researches on vapor and gas phases in which special methods are used for the analysis of the effusing vapors (see the review article by Inghram and Drowart, 1959).

We denote the dimer evaporation coefficient from the bottom and from the walls of the cylinder by δ_a and δ_b. Remembering the definitions of these quantities and the equation for the equilibrium constant

$$K_p = \frac{p_M^2}{p_D},$$

we find that from a plane surface the monomer and dimer evaporation rates are in the ratio $K_p \alpha M^{\frac{1}{2}}/24.17\delta_a \mu_{KM} T^{\frac{1}{2}}$ on the bottom and $K_p \beta M^{\frac{1}{2}}/24.17\delta_b \mu_{KM} T^{\frac{1}{2}}$ on the wall of the cylinder, where μ_{KM} is the effusion rate of monomers from an ideal Knudsen cell in g cm^{-2} sec^{-1}.

If some reverse flow of particles to the evaporating surface is possible, then a number of other effects are superposed on the process of evaporation: condensation, reflection, and other forms of interaction between the vapor and the walls.

In what follows we assume that the ratio of dissimilar particles evaporating from the surface remains the same regardless of the amount of flow toward the surface.

We consider the phenomena in a cylinder, where the magnitude of the flux of particles to a point z on the inside surface of the cylinder depends on the distance from this point to the bottom of the cylinder (it is obvious that there is axial symmetry in this case). If $n_M(z)$ monomers fall on the point z, the fraction ρ_M is reflected and the rest are condensed, whence it is possible for them to evaporate either in the form of dimers or of monomers, with the same ratio of the number of monomers to the number of dimers (see above).

Let $n_D(z)$ dimers fall on the inner surface of the cylinder. Since the problem is being solved for sufficiently low pressures (so that the particles collide more often with the walls and the bottom than with each other), the existing dimer particles will dissociate, insofar as they do so at all, mainly in collisions with the walls.

It is important to note that the Knudsen cosine law is assumed to be valid for all particles coming from any point of the inside surface of the cylinder; in particular this means that the monomers which arrive through dissociation of a dimer at the wall also obey this law.

Thus the fraction ρ_D of the n_D incident dimers which is reflected as dimers at the bottom is ρ_{aDD}, and the corresponding fraction for reflection at a wall is ρ_{bDD},* the fraction ρ'_{DM} is reflected with dissociation into two monomers, and the fraction ρ''_{DM} is "half" reflected (this means that it is possible for one of the monomers which composes a dimer to be condensed and the other to be reflected as a monomer). Condensed dimers, or condensed parts of dimers, behave in just the same way as the condensed monomers (see above).

It cannot be overlooked that some other mechanism might occur in actual fact, but we have tried to take all possibilities into account. If, on the other hand, the phenomenon is simpler, then we can describe it in the same terms by putting some supplementary conditions on the allowed values of the coefficients. For example, if some of the processes we have assumed do not occur (for example, if there is no "half" condensation), then the condition $\rho'' = 0$ holds, and so on.

Accordingly, there will be evaporated and reflected particles coming from a point z, $\Phi_M(z)$ of them being monomers and $\Phi_D(z)$ dimers, per square centimeter per second. We have

$$\Phi_M = n_M \rho_M + 2n_D \rho'_{DM} + n_D \rho''_{DM} + a_1((b_1)).^{**} \tag{12.4}$$

*The symbols α and a are used throughout to describe the behavior of particles at the bottom, and β and b at the wall; if these indices are omitted, the effect holds both for the bottom and for the wall.

**The symbol "a ((b))" means "a or b."

$$\Phi_D = n_D \rho_D + a_2 ((b_2)), \tag{12.5}$$

where a_1 or b_1 is the same as for pure monomer evaporation, and a_2 or b_2 is the rate of evaporation of dimers from the bottom or from the walls.

If, for example, there is reflection of dimers without dissociation, we set $\rho_{DD} = 0$ and get instead of Eq. (12.5) the simpler relation:

$$\Phi_D = a_2 ((b_2)). \tag{12.6}$$

This applies to any part of the equation (12.4). The true behavior of a particular system can evidently be judged by comparing the equations so derived with the experimental results. Since we are not taking collisions between particles into account, on the assumption that the pressure is sufficiently low and that collisions with the bottom and the walls are more frequent, the degree of dissociation of the dimers will increase as the number of the latter collisions increases. To allow for the number of collisions is to allow for the geometry of the vessel from which evaporation is occurring. We shall derive equations which describe the influence of the geometry; first we take the equations for the flow of dimers.

The expression for the number of dimers, analogous to the expression for a vapor of pure monomer composition, will be (cm^{-2}sec^{-1}):

$$-\frac{a_2}{4}\frac{dF}{dz}. \tag{12.7}$$

The number of dimers φ_{D_0} which leave the point z after the first reflection will be (cm^{-2}sec^{-1}):

$$\varphi_{D_0} = b_2 - \rho_{bDD} a_2 F'/4. \tag{12.8}$$

The number of dimers ν_{Di} which strike the bottom after the redistribution of φ_{D_0} will be (cm^{-2} sec^{-1}):

$$\nu_{D_1} = -\frac{1}{2}\int_0^l \varphi_{D_0} F'\, dz. \tag{12.9}$$

Of these particles, just as in the case of evaporation from the bottom, a fraction ρ_{aDD} is reflected in the form of dimers and is again distributed onto the walls of the cylinder according to the formula

$$-\frac{\nu_{D1}\, dF(z)}{4dz}. \tag{12.10}$$

The number of dimers per square centimeter per second impinging on the exit aperture is

$$\frac{\nu_D \rho_{aDD} F(l)}{2} \text{ dimers}\cdot\text{cm}^{-2}\cdot\text{sec}^{-1}.$$

When we take into account the reflection from the walls of particles coming from the walls of the cylinder, as we did in Eq. (4.7') for monomers, we write

$$\varphi_{D_1}(z_1) = \frac{\rho_{bDD}}{4}\int_0^{z_1} \varphi_{D(i-1)} F''(z_1 - z)\, dz +$$

$$+\frac{\rho_{b\,DD}}{4}\int_{z_1}^{l}\varphi_{D(i-1)}F''(z-z_1)\,dz+\frac{\rho_{a\,DD}\rho_{b\,DD}}{8}F'(z_1)\int_0^{l}\varphi_{D(i-1)}F'\,dz \qquad (12.11)$$

and finally the expression for the number of dimers emerging from the point z is

$$\Phi_D(z)=\Sigma\varphi_{Di}$$

$$=\varphi_{D0}+\frac{\rho_{b\,DD}}{4}\int_0^{z_1}\Phi_D F''(z_1-z)\,dz+\frac{\rho_{b\,DD}}{4}\int_{z_1}^{l}\Phi_D F''(z-z_1)\,dz$$

$$+\frac{\rho_{b\,DD}\rho_{b\,DD}}{8}F'(z_1)\int_0^{l}\Phi_D F'(z)\,dz. \qquad (12.12)$$

Equation (12.12) is a formula for the lengthwise distribution of dimers along the cylinder. It can be seen that this expression is identical with the expression (4.8) for monomer particles, and therefore all of the expressions which come from Eq. (4.8) can be converted into the analogous formulas for dimers. We need only replace the coefficients $(1-\alpha)$ and $(1-\beta)$ by the corresponding quantities ρ_{aDD} and ρ_{bDD} for the dimer case, and α and β by the quantities δ_a and δ_b.

For example, we have the following expression:

1) Rate of effusion of dimers from the cylinder (in units of saturated vapor):

$$\bar{\mu}_D(l)=\frac{2\delta_a\delta_b-4\delta_b-2\delta_a\sqrt{\delta_b}+e^{-2l\sqrt{\delta_b}}(4\delta_b-2\delta_a\delta_b-2\delta_a\sqrt{\delta_b})}{\rho_{a\,DD}\rho_{b\,DD}-(1+\sqrt{\delta_b})^2-[(1-\sqrt{\delta_b})^2-\rho_{a\,DD}\rho_{b\,DD}]\,e^{-2l\sqrt{\delta_b}}}, \qquad (12.13)$$

2) Rate of evaporation from the bottom of the cylinder (in units of saturated vapor):

$$\bar{\mu}_D^{b}=1+\frac{4\rho_{a\,DD}\sqrt{\delta_b}\,e^{-l\sqrt{\delta_b}}}{\rho_{a\,DD}\rho_{b\,DD}-(1+\sqrt{\delta_b})^2+[(1-\sqrt{\delta_b})^2-\rho_{a\,DD}\rho_{b\,DD}]\,e^{-2l\sqrt{\delta_b}}}. \qquad (12.14)$$

We shall not translate all of the formulas which were derived in the section on monomers into terms of dimers, but emphasize that this is an entirely legimate and easily performed operation.

Accordingly, the problem of calculating the number of dimers at various horizontal levels in the cylinder, and that of calculating the number of dimers emerging from the cylinder, and so on, is solved. The situation is much more complicated with regard to the calculation of the number of monomers if there are dimers present in the vapor.

We shall derive the equation for the flow of monomers (in the presence of dimers) and take Eq. (5.4) as the starting point. In the derivation of the integral expressions for the monomers (in the presence of dimers) we allow for these facts:

1) that along with the flow of monomers that occur independently and are moving through the cylinder we must include the flow of monomers that originate from dimers;

2) that monomers which have just been produced from dimers at once begin to move according to the laws of monomers;

3) that there are fluxes of both monomers and dimers from the walls to the bottom and the walls, and from the bottom to the walls.

The integral equation for the monomers is of the following form:

$$\Phi=\Phi_M^{I}+\Phi_M^{II}+\Phi_{DM}, \tag{12.15}$$

$$\Phi_M^{I}=b_1-\frac{a_1(1-\beta)F'}{4}+\frac{1-\beta}{4}\int_0^{z_1}\Phi F''(z_1-z)\,dz$$

$$+\frac{1-\beta}{4}\int_{z_1}^{l}\Phi F''(z-z_1)\,dz+\frac{(1-\alpha)(1-\beta)}{8}F'\int_0^{l}\Phi F'\,dz, \tag{12.16}$$

$$\Phi_M^{II}=\frac{2\rho'_{bDM}+\rho''_{bDM}}{4}\left[\int_0^{z_1}\Phi_D F''(z_1-z)\,dz+\int_{z_1}^{l}\Phi_D F''(z-z_1)\,dz\right.$$

$$\left.+\frac{\rho_{aD}}{2}F'\int_0^{l}\Phi_D F'\,dz\right]+\frac{2\rho'_{aDM}+\rho''_{bDM}}{8}(1-\beta)F'\int_0^{l}\Phi_D F'\,dz, \tag{12.17}$$

$$\Phi_{DM}=-\frac{a_2(2\rho'_{bDM}+\rho''_{bDM})F'}{4}+\frac{1-\beta}{4}\int_0^{z_1}\Phi_{DM}F''(z_1-z)\,dz$$

$$+\frac{1-\beta}{4}\int_{z_1}^{l}\Phi_{DM}F''(z-z_1)\,dz+\frac{(1-\alpha)(1-\beta)}{8}F'\int_0^{l}\Phi_{DM}F'\,dz. \tag{12.18}$$

The equation is complicated but can be solved. Nevertheless it is simpler and more correct to make all of the numerical calculations for equations which are special cases of this equation and present less difficulty in the calculations.

A case of practical interest is the problem of the evaporation of a metal from a cylindrical crucible, when neither monomers nor dimers evaporate from the walls of the crucible (in our notation $b_1 = 0$, $b_2 = 0$) Physically it is difficult to imagine that the reflection of monomers would not be close to 1 in this case, since otherwise the mass balance would be upset (in the case of a gas one can assume adsorption-diffusion loss through the wall, but in the vapor case this process is important only in the first moments of the evaporation). We also assume that the interaction of dimers with the wall reduces to the reflection with dissociation and without dissociation of dimers into monomers.

We explain this last remark with the equation of mass balance at the walls:

$$\rho_{bDD}+\rho'_{bDM}=1. \tag{12.19}$$

The equation $\rho''_{bDM}=0$ is due to the fact that ρ'_{bDM} and ρ''_{bDM} occur in the same way in all of the equations and we have no way of distinguishing between them.

Under the assumptions we have made, the terms in Eq. (12.15) can be written as follows:

$$\Phi^{I}_{M}=A+Bz, \tag{12.20}$$

$$\Phi^{II}_{M}=\frac{\rho_{bDM}}{2}\left[\int_0^{z_1}\Phi_D F''(z_1-z)\,dz+\int_{z_1}^{l}\Phi_D F''(z-z_1)\,dz\right]$$

$$+\frac{\rho_{aDD}}{2}F'\int_0^{l}\Phi_D F'dz, \tag{12.21}$$

$$\Phi_{DM}=-\frac{a_2\rho'_{bDM}F'}{2}. \tag{12.22}$$

Equation (12.20) is of the same form as Eqs. (7.5), (7.6):

$$\Phi_D=A_De^{-z\sqrt{1-\rho_{bDD}}}+B_De^{z\sqrt{1-\rho_{bDD}}}$$

$$A_D=-\frac{a_2\rho_{bDD}(1+\sqrt{1-\rho_{bDD}})}{\rho_{aDD}\rho_{bDD}-(1+\sqrt{1-\rho_{bDD}})^2-[\rho_{aDD}\rho_{bDD}-(1-\sqrt{1-\rho_{bDD}})^2]\,e^{-2l\sqrt{1-\rho_{bDD}}}}$$

$$B_D=-\frac{a_2\rho_{bDD}(1-\sqrt{1-\rho_{bDD}})\,e^{-2l\sqrt{1-\rho_{bDD}}}}{\rho_{aDD}\rho_{bDD}-(1+\sqrt{1-\rho_{bDD}})^2-[\rho_{aDD}\rho_{bDD}-(1-\sqrt{1-\rho_{bDD}})^2]\,e^{-2l\sqrt{1-\rho_{bDD}}}}$$

Making the necessary calculations, we get the equation for the distribution of monomers in the horizontal levels of the cylinder:

$$\Phi_M=\frac{a_1(l+1)}{2+\alpha l}-\frac{a_1 z}{2+\alpha l}+A'e^{-z\sqrt{1-\rho_{bDD}}}+B'e^{z\sqrt{1-\rho_{bDD}}}, \tag{12.23}$$

where

$$A'=-\frac{2a_2\rho'_{bDM}[1+\sqrt{1-\rho_{bDD}}]}{\rho_{aDD}\rho_{bDD}-(1+\sqrt{1-\rho_{bDD}})^2-[\rho_{aDD}\rho_{bDD}-(1-\sqrt{1-\rho_{bDD}})^2]\,e^{-2l\sqrt{1-\rho_{bDD}}}}.$$

$$B'=\frac{2a_2\rho_{bDM}(1-\sqrt{1-\rho_{bDD}})\,e^{-2l\sqrt{1-\rho_{bDD}}}}{\rho_{aDD}\rho_{bDD}-(1+\sqrt{1-\rho_{bDD}})^2-[\rho_{aDD}\rho_{bDD}-(1-\sqrt{1-\rho_{bDD}})^2]\,e^{-2l\sqrt{1-\rho_{bDD}}}}.$$

With the assumptions we have made, the effusion rate of monomers from the cylinder is described by an equation of the form

$$\mu_M = a_1 e^{-l} + (1-\alpha) e^{-l} \int_0^l \Phi_M e^{-z} dz + e^{-l} \int_0^l \Phi_M e^{z} dz + 2e^{-l} \rho'_{aDM} \int_0^l \Phi_D e^{-z} dz. \tag{12.24}$$

After substituting the Eq. (12.23) in Eq. (12.24) and carrying out the indicated operations, we get

$$\begin{aligned}\mu_M &= \frac{2a_1}{2+\alpha l} \\ &+ \Big[2a_2 e^{-l}\{[(1-\alpha)(1-\rho_{bDD})\rho_{bDD} + \rho_{aDM}\rho_{bDD}](e^{-2l\sqrt{1-\rho_{bDD}}} - 1) \\ &+ 2a_2 e^{-l}(1-\rho_{bDD})[(1+\sqrt{1-\rho_{bDD}})^2 - 4\sqrt{1-\rho_{bDD}}\, e^{l(1-\sqrt{1-\rho_{bDD}}} \\ &-(1-\sqrt{1-\rho_{bDD}})^2 e^{-2l\sqrt{1-\rho_{bDD}}}]\}\Big]\Big/\rho_{bDD}\{\rho_{aDD}\rho_{bDD} \\ &-(1+\sqrt{1-\rho_{bDD}})^2 - [\rho_{aDD}\rho_{bDD} - (1-\sqrt{1-\rho_{bDD}})^2]\, e^{-2l\sqrt{1-\rho_{bDD}}}\}.\end{aligned} \tag{12.25}$$

Finally, we determine the expression for the evaporation rate of the bottom of the cylinder, assuming that

$$b_1 = b_2 = 0, \tag{12.26}$$

$$2\rho'_{DM} + \rho''_{DM} = 2\rho'_{DM}, \tag{12.27}$$

$$\rho_{bDD} + \rho'_{bDM} = 1; \tag{12.28}$$

we get the integral expression

$$\mu_M^b = a_1 + (1-\alpha) \int_0^l \Phi_M e^{-z} dz + 2\rho'_{aDM} \int_0^l \Phi_D e^{-z} dz. \tag{12.29}$$

The source of the last term in this equation is the **dissociation of dimers into monomers in collisions with the wall.**

After carrying out the necessary operations, we get

$$\mu_M^b = a_1 \left[1 + \frac{(1-\alpha) l}{2+\alpha l}\right] + \frac{2a_2 (e^{-2l\sqrt{1-\rho_{bDD}}} - 1)[(1-\alpha)(1-\rho_{bDD}) + \rho'_{bDM}\rho_{bDD}]}{\rho_{aDD}\rho_{bDD} - (1+\sqrt{1-\rho_{bDD}})^2 - [\rho_{aDD}\rho_{bDD} - (1-\sqrt{1-\rho_{bDD}})^2]\, e^{-2l\sqrt{1-\rho_{bDD}}}} \tag{12.30}$$

The Clausing coefficient in the usual sense corresponds to a certain resistance which the tube offers to the molecular flow along it owing to collisions of particles with the walls of the tube and the scattering of particles according to the Knudsen cosine law. When, however, the molecular flow along the tube consists of particles which can be destroyed in collisions with the walls, it is also necessary to take this interaction into account.

If in every collision – for example between a dimer and the wall of the tubes – there is a probability of destruction equal to $(1-\rho_{bDD})$, then we calculate the Clausing coefficient W_x of a tube of unit radius and length l from the formula

$$W_x = \frac{4\sqrt{1-\rho_{bDD}}\, e^{l\sqrt{1-\rho_{bDD}}}}{(1+\sqrt{1-\rho_{bDD}})^2\, e^{2l\sqrt{1-\rho_{bDD}}} - (1-\sqrt{1-\rho_{bDD}})^2}. \qquad (12.31)$$

For $\rho_{bDD}=1$, Eq. (12.31) gives the formula obtained earlier by Clausing. The equation represents a possible generalization of Clausing's result.

The relations we have derived can also be applied to other reactions in which particles are disintegrated at the walls of channels, if appropriate assumptions are made.

SECTION 13

THE TABULATION OF CERTAIN RELATIONS

The relations we have derived are often cumbersome and not very convenient to use in work on the determination of vapor pressures, rates of evaporation, evaporation coefficients, and so on. Therefore, we have resorted to the possibilities afforded by modern computing techniques, and have made a number of calculations with the high-speed electronic computer "Strela" at the Computing Center of the Academy of Sciences of the USSR.

The amount of information which can ordinarily be obtained with an apparatus – for example, a mass spectrometer – is finite, and the data which can be read off from the device are discrete. The first fact means that the device measures only in a definite range of values. For example, a mass spectrometer with an electrometer amplifier measures ion currents from 10^{-9} to 10^{-14} A, and when a multiplier is used, from 10^{-12} to 10^{-17} A, i.e., over a range of four or five orders of magnitude. Ordinary analytical balances also measure over a range of five or six orders of magnitude. The second fact means that the device has a definite measurement error. Therefore, the numerical data obtained with it form a discrete series of values. Suppose that a mass spectrometer has an accuracy of two percent; then the ratio of its readings in any two cases will not have an error worse than four percent. We assume that we can distinguish ratios of the readings which differ by not less than this amount of four percent. It is more convenient to consider ratios of readings in order to get dimensionless numbers. This means that the conclusions and results we get at the end of our calculations do not depend on the type of device from which we are taking the readings. In other words, we can equally well apply our calculations either to a mass spectrometer solution of a problem about evaporation or to any other procedure dealing in terms of masses, and also to various methods of studying evaporation with radioactive isotopes, or to possible combinations of these two sorts of methods.

It is easily seen that with a device which measures over a range of five orders of magnitude with an accuracy of two percent the ratios of any of its readings (smaller reading divided by larger reading) will differ by four percent; that is, a series of numbers $\overline{\mu}_i$ is defined which is a geometric progression whose first term is 10^{-5}, whose ratio is 1.04, and whose last term is 1. A simple calculation shows that there are 294 terms in this series, and this number fixes the amount of information which can be obtained from the device with certainty. If the accuracy of the measurements is increased it is easy to make an interpolation between adjacent values.

Accordingly, when there is a change of the rate of evaporation owing to a change of the effusion coefficient, and this is measured with some particular device, we get the ratio of the two readings in the form of a number belonging to a series of numbers that we have already ascertained.

Next, using the relations which have been derived (between the evaporation rate, the effusion coefficient, the evaporation coefficient, the ratio l of the length of the cylinder to its radius, and so on), and having measured the rates of evaporation from two cylinders with different values of l, we can obtain the other unknown if we have ready-made solutions of certain transcendental equations in tabulated form.

We can make these general remarks clearer by considering the tables which we have obtained and some examples of their use.

The general formula (7.22) which gives the rate of effusion from the cylinder has been considered in Section 7. If we are interested in the flow out of a cylinder (Fig. 8) which consists entirely of the material under investigation, then Eq. (7.22) can be written in the form

$$\mu = \frac{a\,[(2\alpha^2 - 4\alpha - 2\alpha\sqrt{\alpha}) + (4\alpha - 2\alpha^2 - 2\alpha\sqrt{\alpha})\,e^{-2l\sqrt{\alpha}}]}{\alpha\,\{(1-\alpha)^2 - (1+\sqrt{\alpha})^2 - [(1-\alpha)^2 - (1-\sqrt{\alpha})^2]\,e^{-2l\sqrt{\alpha}}\}}. \qquad (13.1)$$

We rewrite Eq. (13.1) in the form

$$\frac{a}{\mu}=\frac{(1-\alpha)^2-(1+\sqrt{\alpha})^2-[(1-\alpha)^2-(1-\sqrt{\alpha})^2]\,e^{-2l\sqrt{\alpha}}}{2\alpha-4-2\sqrt{\alpha}+(4-2\alpha-2\sqrt{\alpha})\,e^{-2l\sqrt{l}}}. \tag{13.2}$$

Here the left-hand term is the ratio of the evaporation rate as defined by Langmuir to the evaporation rate from a cylinder with certain values of l and α.

As has been indicated, one can determine a and μ (from the loss of weight, by using radioactive or stable isotopes, and so on). As has been stated, the ratio a^i/μ_i will be one of the terms of a geometric progression whose first term is 10^{-5}, whose ratio is 1.04, and whose last term is 1.

If we substitute the terms of this series successively in the left-hand term of Eq. (13.2) and fix a particular value of l (for example, 4, 8, 12), then by solving the resulting equation for α we get a table of the possible values of

$$\alpha=f(\mu,\ l).$$

The calculations were run on the computer "Strela" and the results are given in Table I.

Having obtained a ratio a/μ which appears in the left-hand column of Table I, we can determine α if we carry out evaporations first with a flat specimen and then from cylinders with $l=4$, 8, 12.

Knowing a and α, we can easily determine the saturated vapor pressure p from the formula

$$p=\frac{a}{\alpha}\sqrt{\frac{2\pi RT}{M}}. \tag{13.3}$$

Obviously the effusion coefficients of cylinders made of the material in question can also be determined by means of Table I. Namely: knowing a, μ, and α we can easily determine $\mu\alpha/a$, which is the effusion coefficient,* but it is helpful to give a table of the values of $\mu\alpha/a$ as a function of α and l.

We rewrite Eq. (13.1) in the form

$$\frac{\mu\alpha}{a}=\frac{2\alpha^2-4\alpha-2\alpha\sqrt{\alpha}+(4\alpha-2\alpha^2-2\alpha\sqrt{\alpha})\,e^{-2l\sqrt{l}}}{(1-\alpha)^2-(1+\sqrt{\alpha})^2-[(1-\alpha^2)-(1-\sqrt{\alpha})^2]\,e^{-2l\sqrt{\alpha}}}. \tag{13.4}$$

We now do the same thing that we did with Eq. (13.2), and get Table II. A possible experiment, working with this table, consists in comparing the rates of evaporation from cylinders with definite values of l and some (as yet unknown) value of α with rates from an ideal Knudsen cell.

To determine how close the vapor near the bottom of a given cylinder is to saturation, we use Eq. (7.12), which is tabulated in Table III. Table III gives for various values of l the dependence of the degree of saturation μ_M^b of the vapor at the bottom of a cylinder on the value α.

It should be pointed out that if we take α to mean the specific emissive power (spectral or monochromatic) of the material of the cylinder, than Tables I, II, III enable us to calculate a number of photometric characteristics, and to do so we must measure instead of the evaporation rate a and effusion rate μ the total or monochromatic radiant emittances of the cylinder and of the material of which it is made.

*The effusion coefficient of a vessel of given shape is the ratio of the rate of effusion from this vessel to the rate of effusion from an ideal Knudsen cell.

To determine the reflection coefficients of dimers from the walls of a cylinder in the simple case when the bottom and the walls of the cylinder are made of the same material (the one under investigation), we can use Eq. (13.2) and Table I, replacing α by $(1-\rho_{aDD})$ to obtain the desired solution

The analogous problem for monomers is simple, since we can set $\beta = 0$ in the corresponding equations; then the equations take a rather simple form and there is no need to use machine calculations.

A more complicated problem is that of finding coefficients of the type ρ_{bDD} – the coefficient of "nondestructive" reflection of dimers from the walls – for example, those of a ceramic crucible, at the bottom of which is the condensed phase from which the dimers evaporate.

It is logical to assume that the walls do not evaporate the dimers. Accordingly, in Eq. (13.5), which describes the evaporation rate of dimers from a cylinder,

$$\mu_D = \frac{b_2\left[(2\delta_a\delta_b - 4\delta_b - 2\delta_a\sqrt{\delta_b}\,) + (4\delta_b - 2\delta_a\delta_b - 2\delta_a\sqrt{\delta_b}\,)\,e^{-2l\sqrt{\delta_b}}\right] + 4\frac{\sqrt{\delta_b}}{b_2}\,\delta_a(\delta_a b_2 - \delta_b a_2)\,e^{-l\sqrt{\delta_b}}}{\delta_b(1-\delta_a)(1-\delta_b) - (1+\sqrt{\delta_b}\,)^2 - \left[(1-\delta_a)(1-\delta_b) - (1-\sqrt{\delta_b}\,)^2\right]e^{-2l\sqrt{\delta_b}}}; \qquad (13.5)$$

We can assume that $b_2 = 0$; we then get

$$\frac{\mu_D}{a_2} = \frac{4\sqrt{1-\rho_{bDD}}\,e^{-l\sqrt{1-\rho_{bDD}}}}{\rho_{aDD}\rho_{bDD} - (1+\sqrt{1-\rho_{bDD}}\,)^2 - \left[\rho_{aDD}\rho_{bDD} - (1-\sqrt{1-\rho_{bDD}}\,)^2\right]e^{-2l\sqrt{1-\rho_{bDD}}}} \qquad (13.6)$$

Equation (13.6) has been tabulated on the same principle as Eqs. (13.2) and (13.4). Table IV shows the dependence of ρ_{bDD} on the various values of μ_{Di}/a_{2i}, ρ_{aDD}, l. The values of ρ_{aDD} must be determined from experiments with a cylinder made of the material to be investigated. The values of μ_{Di}/a_{2i} lie in the same ranges as those of a_i/μ_i and $\mu_i\alpha/a_{1i}$ in Tables I and II, respectively. The values of ρ_{aDD} and l at which ρ_{bDD} was calculated are shown in the table.

Equation (12.31) shows to what extent the flux of dimer particles is attenuated on going through a tube for which the ratio of length to radius is l, if the probability of dissociation of dimer particles into monomers is $(1-\rho_{bDD})$.

Evidently this same formula also correctly describes other processes of dissociation of the particles of a vapor phase at the wall of the vessel, if the frequency of collisions of particles with the wall is several orders of magnitude larger than the frequency of collisions between particles.

Table V is calculated from Eq. (12.31):

$$W_x = \frac{4\sqrt{1-\rho_{bDD}}\,.\,e^{l\sqrt{1-\rho_{bDD}}}}{(1+\sqrt{1-\rho_{bDD}}\,)^2\,e^{2l\sqrt{1-\rho_{bDD}}} - (1-\sqrt{1-\rho_{bDD}}\,)^2}. \qquad (13.7)$$

By using Table V we can determine the probability of dissociation of a particular type of particles colliding with the walls of a tube.

Let the concentration of dimer or other particles at the entrance to a particular tube be c_1, and that at the exit c_2. Then by determining

$$W_x = \frac{c_2}{c_1}$$

and using Table V it is easy to find ρ_{bDD} for tubes with values of l which vary* from 0.1 to 30, and values of $1-\rho_{bDD}$ which vary from 10^{-4} to 10^{-2}.

*Concerning data for $l > 8$ see footnote on page 41.

Table I

$\frac{a}{\mu}$	α: $l = 4$	$= 8$	$= 12$	$\frac{a}{\mu}$	α: $l = 4$	$= 8$	$= 12$	$\frac{a}{\mu}$	α: $l = 4$	$= 8$	$= 12$
$3759 \cdot 10^{-5}$	—	—	—	$1219 \cdot 10^{-4}$	$8703 \cdot 10^{-6}$	$3675 \cdot 10^{-5}$	$4030 \cdot 10^{-5}$	$3950 \cdot 10^{-4}$	$2655 \cdot 10^{-4}$	$2701 \cdot 10^{-4}$	$2702 \cdot 10^{-4}$
3910	—	—	—	1267	$1269 \cdot 10^{-5}$	3991	4323	4108	2822	2863	2864
4066	—	—	$2559 \cdot 10^{-7}$	1318	1686	4323	4634	4272	2998	3034	3035
4228	—	—	8914	1371	2122	4675	4964	4443	3182	3215	3215
4398	—	—	$1559 \cdot 10^{-6}$	1425	2578	5045	5313	4620	3376	3405	3405
4573	—	—	2262	1482	3056	5436	5684	4805	3580	3605	3605
4756	—	—	3000	1542	3557	5849	6078	4997	3794	3816	3816
4946	—	—	3778	1603	4081	6285	6495	5197	4019	4039	4039
5144	—	—	4596	1667	4630	6746	6937	5405	4256	4273	4273
5350	—	—	5458	1734	5205	7233	7406	5621	4505	4519	4519
5564	—	—	6365	1803	5808	7747	7904	5845	4766	4778	4778
5786	—	—	7322	1875	6439	8290	8432	6079	5041	5051	5051
6017	—	$7068 \cdot 10^{-7}$	8330	1950	7101	8864	8991	6322	5330	5338	5338
6258	—	$1975 \cdot 10^{-6}$	9393	2028	7795	9470	9584	6575	5633	5640	5640
6508	—	3306	$1051 \cdot 10^{-5}$	2110	8523	$1011 \cdot 10^{-4}$	$1021 \cdot 10^{-4}$	6838	5952	5957	5957
6768	—	4703	1170	2194	9286	1079	1088	7111	6286	6291	6291
7039	—	6170	1295	2282	$1009 \cdot 10^{-4}$	1150	1158	7395	6638	6641	6641
7320	—	7710	1427	2373	1093	1226	1233	7691	7007	7010	7010
7613	—	9328	1566	2468	1181	1306	1312	7999	7395	7397	7397
7917	—	$1102 \cdot 10^{-5}$	1714	2566	1273	1391	1396	8318	7802	7803	7803
8233	—	1282	1870	2669	1371	1480	1484	8651	8229	8230	8230
8563	—	1470	2034	2776	1472	1574	1578	8997	8677	8678	8678
8905	—	1667	2209	2887	1580	1674	1677	9356	9148	9148	9148
9261	—	1875	2393	3002	1692	1779	1782	9730	9642	9642	9642
9632	—	2094	2589	3122	1810	1891	1893				
$1002 \cdot 10^{-4}$	—	2324	2795	3247	1934	2008	2010				
1042	—	2567	3015	3377	2064	2132	2134				
1083	—	2823	3247	3512	2201	2263	2264				
1127	$1253 \cdot 10^{-6}$	3092	3493	3652	2345	2401	2402				
1172	4894	3376	3754	3798	2496	2547	2548				

Table II

$\frac{\mu\alpha}{a}$	α $l = 4$	$= 8$	$= 12$	$\frac{\mu\alpha}{a}$	α $l = 4$	$= 8$	$= 12$	$\frac{\mu\alpha}{a}$	α $l = 4$	$= 8$	$= 12$
$1000 \cdot 10^{-8}$	$1113 \cdot 10^{-9}$	$5882 \cdot 10^{-10}$	$3974 \cdot 10^{-10}$	$5188 \cdot 10^{-8}$	$5762 \cdot 10^{-9}$	$3056 \cdot 10^{-9}$	$2078 \cdot 10^{-9}$	$2692 \cdot 10^{-7}$	$2991 \cdot 10^{-8}$	$1585 \cdot 10^{-8}$	$1078 \cdot 10^{-8}$
1040	1160	6120	4213	5396	6000	3175	2162	2799	3112	1648	1121
1082	1196	6358	4332	5611	6239	3306	2245	2911	3236	1713	1165
1125	1256	6597	4456	5836	6489	3437	2341	3028	3365	1783	1212
1170	1303	6835	4690	6069	6739	3568	2424	3149	3500	1853	1260
1217	1351	7193	4809	6312	7014	3711	2519	3275	3640	1928	1311
1265	1411	7431	5047	6564	7300	3866	2627	3406	3785	2004	1363
1316	1458	7789	5286	6826	7586	4021	2734	3542	3937	2085	1418
1368	1518	8027	5524	7099	7883	4176	2841	3683	4094	2168	1475
1423	1578	8385	5643	7383	8206	4343	2949	3831	4259	2255	1534
1480	1649	8743	5882	7678	8528	4522	3068	3984	4429	2345	1595
1539	1709	9100	6120	7985	8873	4701	3199	4143	4605	2439	1659
1601	1780	9458	6358	8305	9231	4892	3318	4309	4790	2536	1725
1665	1851	9816	6716	8637	9600	5082	3461	4481	4981	2639	1795
1731	1923	$1017 \cdot 10^{-9}$	6955	8982	9982	5285	3592	4660	5181	2743	1866
1800	1995	1065	7193	9341	$1038 \cdot 10^{-8}$	5500	3735	4846	5387	2853	1941
1872	2078	1101	7431	9714	1079	5714	3890	5040	5603	2968	2019
1947	2162	1148	7789	$1010 \cdot 10^{-7}$	1122	5941	4045	5242	5872	3087	2091
2025	2245	1196	8147	1051	1167	6179	4200	5451	6061	3210	2183
2106	2341	1244	8385	1093	1214	6429	4367	5669	6303	3338	2271
2190	2436	1291	8743	1136	1263	6692	4546	5896	6556	3472	2362
2278	2531	1339	9100	1182	1313	6954	4725	6131	6818	3611	2456
2366	2627	1399	9458	1229	1365	7228	4915	6376	7091	3755	2555
2464	2734	1446	9816	1278	1420	7526	5118	6631	7375	3906	2658
2562	2841	1506	$1029 \cdot 10^{-9}$	1329	1477	7824	5321	6897	7669	4062	2764
2664	2960	1566	1065	1382	1536	8134	5535	7172	7977	4225	2875
2771	3080	1625	1113	1438	1598	8456	5750	7459	8295	4394	2989
2882	3199	1697	1148	1495	1661	8802	5988	7757	8628	4570	3109
2997	3330	1768	1196	1555	1728	9147	6227	8067	8972	4753	3233
3117	3461	1828	1244	1617	1797	9517	6465	8390	9332	4913	3363
3241	3604	1911	1291	1682	1863	9898	6727	8725	9705	5141	3498
3371	3747	1983	1351	1749	1944	$1029 \cdot 10^{-8}$	7002	9074	$1009 \cdot 10^{-7}$	5347	3639
3506	3890	2066	1399	1819	2021	1070	7276	9437	1050	5562	3784
3646	4045	2150	1458	1892	2102	1113	7574	9814	1092	5783	3936
3792	4212	2233	1518	1967	2187	1158	7878	$1021 \cdot 10^{-6}$	1136	6016	4093
3943	4379	2317	1578	2046	2274	1204	8182	1061	1181	6257	4257
4101	4558	2412	1637	2128	2364	1252	8516	1104	1228	6507	4428
4265	4737	2507	1709	2213	2460	1302	8861	1148	1277	6768	4605
4435	4927	2615	1780	2301	2557	1355	9207	1194	1329	7039	4790
4613	5130	2710	1831	2393	2660	1408	9577	1242	1382	7321	4982
4797	5333	2817	1923	2489	2766	1464	9958	1291	1437	7614	5182
4989	5547	2937	1995	2588	2877	1524	$1036 \cdot 10^{-8}$	1343	1495	7919	5390

Table II (cont.)

$\frac{\mu\alpha}{a}$	α			$\frac{\mu\alpha}{a}$	α			$\frac{\mu\alpha}{a}$	α		
	$l = 4$	$= 8$	$= 12$		$l = 4$	$= 8$	$= 12$		$l = 4$	$= 8$	$= 12$
$1397 \cdot 10^{-6}$	$1554 \cdot 10^{-7}$	$8238 \cdot 10^{-8}$	$5607 \cdot 10^{-8}$	$2828 \cdot 10^{-6}$	$3153 \cdot 10^{-7}$	$1673 \cdot 10^{-7}$	$1140 \cdot 10^{-7}$	$5727 \cdot 10^{-6}$	$6409 \cdot 10^{-7}$	$3406 \cdot 10^{-7}$	$2325 \cdot 10^{-7}$
1452	1617	8567	5832	2941	3280	1740	1185	5956	6667	3544	2420
1510	1681	8911	6066	3059	3412	1810	1233	6194	6936	3687	2518
1571	1749	9268	6309	3181	3549	1883	1283	6442	7215	3837	2620
1634	1819	9640	6563	3308	3691	1958	1335	6699	7506	3992	2727
1699	1892	$1003 \cdot 10^{-7}$	6826	3441	3839	2037	1389	6967	7809	4154	2838
1767	1968	1043	7101	3578	3993	2119	1445	7246	8124	4322	2953
1838	2046	1085	7385	3721	4154	2205	1503	7536	8452	4497	3074
1911	2128	1128	7682	3870	4321	2293	1560	7837	8793	4680	3199
1987	2214	1174	7991	4025	4494	2386	1627	8150	9148	4870	3330
2067	2302	1221	8312	4186	4675	2482	1692	8476	9518	5068	3460
2150	2395	1270	8645	4353	4863	2582	1761	8815	9903	5274	3608
2235	2491	1321	8994	4527	5058	2686	1832	9167	$1030 \cdot 10^{-6}$	5489	3755
2325	2591	1374	9335	4708	5262	2794	1906	9534	1072	5712	3909
2418	2694	1429	9731	4896	5473	2907	1983	9915	1115	5945	4070
2514	2803	1486	$1012 \cdot 10^{-7}$	5092	5694	3025	2064	$1031 \cdot 10^{-5}$	1161	6187	4237
2615	2915	1546	1053	5295	5923	3147	2147	1072	1208	6440	4411
2719	3032	1608	1095	5507	6161	3274	2234	1115	1256	6703	4593

Table II (cont.)

$\frac{\mu\alpha}{a}$	α		$\frac{\mu\alpha}{a}$	α		$\frac{\mu\alpha}{a}$	α		$\frac{\mu\alpha}{a}$	α	
	$l = 4$	$= 8$		$l = 4$	$l = 8$		$l = 4$	$l = 8$		$l = 4$	$l = 8$
$1160 \cdot 10^{-5}$	$1307 \cdot 10^{-6}$	$6977 \cdot 10^{-7}$	$2349 \cdot 10^{-5}$	$2688 \cdot 10^{-6}$	$1446 \cdot 10^{-6}$	$4756 \cdot 10^{-5}$	$5615 \cdot 10^{-6}$	$3075 \cdot 10^{-6}$	$9261 \cdot 10^{-5}$	$1162 \cdot 10^{-5}$	$6590 \cdot 10^{-6}$
1206	1361	7262	2443	2799	1507	4946	5854	3210	9632	1214	6910
1254	1416	7560	2540	2914	1570	5144	6104	3352	$1002 \cdot 10^{-4}$	1269	7248
1304	1473	7870	2642	3035	1636	5350	6365	3501	1042	1327	7606
1357	1533	8193	2747	3160	1705	5564	6638	3657	1083	1389	7984
1411	1596	8529	2857	3291	1777	5786	6924	3821	1127	1453	8386
1467	1661	8880	2971	3428	1853	6017	7223	3993	1172	1521	8811
1526	1728	9246	3090	3570	1931	6258	7536	4174	1219	1592	9262
1587	1799	9627	3214	3719	2013	6508	7863	4364	1267	1667	9742
1650	1872	$1002 \cdot 10^{-6}$	3342	3874	2099	6768	8206	4563	1318	1747	$1025 \cdot 10^{-5}$
1716	1949	1044	3476	4036	2189	7039	8565	4773	1371	1830	1079
1785	2028	1087	3615	4205	2283	7320	8941	4994	1425	1919	1137
1856	2111	1132	3759	4381	2381	7613	9336	5226	1482	2012	1199
1931	2198	1179	3910	4565	2484	7917	9749	5471	1542	2111	1265
2008	2288	1228	4066	4757	2592	8233	$1018 \cdot 10^{-5}$	5729	1603	2216	1335
2088	2382	1279	4228	4958	2704	8563	1064	6001	1667	2326	1411
2172	2480	1333	4398	5167	2822	8905	1111	6288	1734	2444	1491
2258	2582	1388	4573	5386	2945						

Table II (cont.)

$\frac{\mu\alpha}{a}$	α		$\frac{\mu\alpha}{a}$	α		$\frac{\mu\alpha}{a}$	α		$\frac{\mu\alpha}{a}$	α	
	$l=4$	$l=8$		$l=4$	$l=8$		$l=4$	$l=8$		$l=4$	$l=8$
$1803\cdot10^{-4}$	$2568\cdot10^{-5}$	$1578\cdot10^{-5}$	$3002\cdot10^{-4}$	$5206\cdot10^{-5}$	$3643\cdot10^{-5}$	$4997\cdot10^{-4}$	$1293\cdot10^{-4}$	$1124\cdot10^{-4}$	$7999\cdot10^{-4}$	$4475\cdot10^{-4}$	$4412\cdot10^{-4}$
1875	2701	1672	3122	5531	3925	5197	1407	1244	8318	5094	5071
1950	2841	1772	3247	5883	4236	5405	1535	1381	8651	5826	5810
2028	2991	1880	3377	6266	4581	5621	1681	1535	8997	6695	6685
2110	3151	1997	3512	6683	4963	5845	1846	1711	9356	7732	7728
2194	3321	2124	3652	7138	5388	6079	2035	1911	9730	8979	8978
2282	3503	2261	3798	7636	5862	6322	2250	2140			
2373	3697	2411	3950	8183	6391	6575	2498	2401			
2468	3906	2573	4108	8784	6983	6838	2784	2700			
2566	4129	2750	4272	9448	7647	1117	3115	3045			
2669	4369	2943	4443	$1018\cdot10^{-4}$	8392	7395	3500	3443			
2776	4627	3155	4620	1100	9230	7691	3948	3905			
2887	4905	3387	4805	1191	$1017\cdot10^{-4}$						

Table III

α	$\bar{\mu}^b_M$				α	$\bar{\mu}^b_M$			
	$l=4$	$=8$	$=12$	$=16$		$l=4$	$=8$	$=12$	$=16$
$1\cdot10^{-6}$	$1499\cdot10^{-8}$	$4499\cdot10^{-8}$	$9098\cdot10^{-8}$	$1530\cdot10^{-7}$	$1\cdot10^{-2}$	$1335\cdot10^{-4}$	$3266\cdot10^{-4}$	$5138\cdot10^{-4}$	$6624\cdot10^{-4}$
3	4498	$1350\cdot10^{-7}$	$2729\cdot10^{-7}$	4588	3	3276	6277	8087	9037
7	$1050\cdot10^{-7}$	3149	6367	$1070\cdot10^{-6}$	7	5583	8396	9440	9806
$1\cdot10^{-5}$	1500	4498	9093	1528	$1\cdot10^{-1}$	6617	9019	9723	9922
3	4498	$1348\cdot10^{-6}$	$2724\cdot10^{-6}$	4573	2	8366	9725	9954	9992
7	$1049\cdot10^{-6}$	3142	6336	$1062\cdot10^{-5}$	3	9102	9899	9989	9999
$1\cdot10^{-4}$	1498	4483	9032	1511	4	9476	9958	9997	10000
3	4483	$1335\cdot10^{-5}$	2669	4421	5	9686	9981	9999	10000
7	$1041\cdot10^{-5}$	3069	$6049\cdot10^{-5}$	9831	6	9813	9992	10000	10000
$1\cdot10^{-3}$	1482	$1213\cdot10^{-4}$	$2222\cdot10^{-4}$	$3312\cdot10^{-4}$	7	9892	9996	10000	10000
7	9663	2493	4144	5607	8	9944	9998	10000	10000
					9	9977	999	10000	10000

Table IV

$\rho_a DD =$	$=0.01$	$=0.1$	$=0.5$	$=0.9$	$=0.99$	$\rho_a DD =$	$=0.01$	$=0.1$	$=0.5$	$=0.9$	$=0.99$
$\frac{\mu_D}{a_s}$	$l=12$	$=12$	$=12$	$=12$	$=12$	$\frac{\mu_D}{a_s}$	$l=12$	$=12$	$=12$	$=12$	$=12$
$1000 \cdot 10^{-8}$	$7957 \cdot 10^{-5}$	$7927 \cdot 10^{-5}$	$7797 \cdot 10^{-5}$	$7670 \cdot 10^{-5}$	$7642 \cdot 10^{-5}$	$3241 \cdot 10^{-8}$	$2587 \cdot 10^{-4}$	$2577 \cdot 10^{-4}$	$2534 \cdot 10^{-4}$	$2491 \cdot 10^{-4}$	$2482 \cdot 10^{-4}$
1040	8583	8551	8411	8273	8243	3371	2643	2633	2589	2546	2536
1082	9208	9173	9023	8875	8843	3506	2700	2688	2644	2600	2590
1125	9830	9794	9632	9475	9440	3646	2756	2745	2699	2654	2644
1170	$1045 \cdot 10^{-4}$	$1041 \cdot 10^{-4}$	$1024 \cdot 10^{-4}$	$1007 \cdot 10^{-4}$	$1004 \cdot 10^{-4}$	3792	2812	2801	2754	2708	2697
1217	1107	1103	1085	1067	1063	3943	2867	2857	2809	2761	2751
1265	1169	1164	1145	1126	1122	4101	2923	2912	2863	2815	2804
1316	1230	1225	1205	1185	1181	4265	2978	2967	2917	2868	2857
1368	1291	1286	1265	1244	1240	4435	3033	3022	2971	2921	2909
1423	1352	1347	1325	1303	1298	4613	3088	3077	3025	2974	2962
1480	1413	1408	1385	1362	1357	4797	3143	3131	3079	3026	3014
1539	1474	1468	1444	1420	1415	4989	3197	3185	3132	3079	3067
1601	1534	1528	1503	1478	1473	5188	3252	3230	3185	3131	3119
1665	1594	1588	1562	1536	1531	5396	3306	3293	3238	3183	3170
1731	1654	1648	1621	1594	1588	5611	3360	3347	3291	3235	3222
1800	1714	1708	1679	1652	1645	5836	3413	3400	3343	3286	3273
1872	1774	1767	1738	1709	1703	6069	3467	3454	3396	3337	3324
1947	1833	1826	1796	1766	1760	6312	3520	3507	3448	3389	3375
2025	1892	1885	1854	1823	1816	6564	3573	3560	3500	3440	3426
2106	1951	1944	1912	1880	1873	6826	3626	3612	3552	3490	3477
2190	2010	2003	1969	1937	1929	7099	3678	3665	3603	3541	3527
2278	2069	2061	2027	1993	1985	7383	3731	3717	3654	3591	3577
2369	2127	2119	2084	2049	2041	7678	3783	3769	3706	3642	3627
2464	2185	2177	2141	2105	2097	7985	3885	3821	3757	3692	3677
2562	2243	2235	2198	2161	2153	8305	3887	3872	3807	3741	3727
2664	2301	2292	2254	2216	2208	8637	3939	3924	3858	3791	3776
2771	2359	2350	2311	2272	2263	8982	3990	3975	3908	3840	3825
2882	2416	2407	2367	2327	2318	9341	4041	4026	3958	3890	3874
2997	2473	2464	2423	2382	2373	9714	4092	4077	4008	3939	3923
3117	2530	2520	2478	2437	2427	$1010 \cdot 10^{-7}$	4143	4128	4058	3987	3971

Table IV (cont.)

$\rho_a DD=$	$=0.01$	$=0.1$	$=0.5$	$=0.9$	$=0.99$	$\rho_a DD=$	$=0.01$		$=0.1$		$=0.5$
$\frac{\mu D}{a_s}$	$l=12$	$=12$	$=12$	$=12$	$=12$	$\frac{\mu D}{a_s}$	$l=8$	$=12$	$=8$	$=12$	$=8$
$1051 \cdot 10^{-7}$	$4194 \cdot 10^{-4}$	$4178 \cdot 10^{-4}$	$4108 \cdot 10^{-4}$	$4036 \cdot 10^{-4}$	$4020 \cdot 10^{-4}$	$3406 \cdot 10^{-7}$	—	$5617 \cdot 10^{-4}$	—	$5596 \cdot 10^{-4}$	—
1093	4244	4228	4157	4084	4068	3542	—	5661	—	5640	—
1136	4294	4278	4206	4133	4116	3683	—	5705	—	5684	—
1182	4344	4328	4255	4181	4164	3831	—	5749	—	5728	—
1229	4394	4378	4304	4229	4211	3984	$4251 \cdot 10^{-5}$	5793	$4227 \cdot 10^{-5}$	5771	$4124 \cdot 10^{-5}$
1278	4444	4427	4352	4276	4259	4143	5208	5836	5179	5814	5053
1329	4493	4476	4401	4324	4306	4309	6161	5879	6127	5857	5977
1382	4542	4525	4449	4371	4353	4481	7110	5922	7070	5900	6897
1438	4591	4574	4497	4418	4400	4660	8054	5965	8008	5943	7813
1495	4640	4623	4544	4465	4446	4846	8993	6008	8943	5986	8724
1555	4689	4671	4592	4511	4493	5040	9928	6050	9872	6028	9631
1617	4737	4719	4639	4558	4539	5242	$1086 \cdot 10^{-4}$	6092	$1080 \cdot 10^{-4}$	6070	$1053 \cdot 10^{-4}$
1682	4785	4767	4686	4604	4585	5451	1178	6134	1172	6112	1143
1749	4833	4815	4733	4650	4631	5669	1271	6176	1263	6153	1232
1819	4881	4862	4780	4696	4676	5896	1362	6218	1355	6195	1321
1892	4928	4910	4827	4741	4722	6131	1454	6259	1445	6236	1410
1967	4976	4957	4873	4787	4767	6376	1544	6300	1536	6277	1498
2046	5023	5004	4919	4832	4812	6631	1635	6341	1625	6318	1585
2128	5070	5051	4965	4877	4857	6897	1725	6382	1715	6359	1673
2213	5116	5097	5011	4922	4902	7172	1814	6423	1804	6399	1759
2301	5163	5144	5056	4967	4946	7459	1903	6463	1892	6439	1846
2393	5209	5190	5102	5011	4990	7757	1992	6503	1980	6479	1931
2489	5255	5236	5147	5055	5034	8067	2080	6543	2068	6519	2017
2588	5301	5281	5192	5099	5078	8390	2167	6583	2155	6559	2102
2692	5347	5327	5237	5143	5122	8725	2254	6622	2241	6598	2186
2799	5392	5372	5281	5187	5165	9074	2341	6662	2328	6637	2270
2911	5438	5417	5326	5230	5208	9437	2428	6701	2414	6676	2354
3028	5483	5462	5370	5273	5251	9814	2513	6740	2498	6715	2437
3149	5528	5507	5414	5317	5294	$1021 \cdot 10^{-6}$	2599	6779	2584	6754	2520
3275	5572	5552	5457	5359	5337	1061	2684	6817	2669	6792	2602

Table IV (cont.)

$\rho_a DD=$	$=0.5$	$=0.9$		$=0.99$	
$\frac{\mu D}{a_s}$	$l=12$	$=8$	$=12$	$=8$	$=12$
$3406\cdot 10^{-7}$	$5501\cdot 10^{-4}$	$3565\cdot 10^{-6}$	$5402\cdot 10^{-4}$	$3546\cdot 10^{-6}$	$5379\cdot 10^{-4}$
3542	5544	$1280\cdot 10^{-5}$	5445	$1274\cdot 10^{-5}$	5422
3683	5587	2200	5487	2188	5464
3831	5630	3115	5529	3099	5505
3984	5673	4026	5571	4004	5547
4143	5716	4932	5612	4906	5589
4309	5758	5834	5654	5803	5630
4481	5800	6732	5695	6696	5671
4660	5842	7625	5736	7584	5712
4846	5884	8514	5777	8468	5752
5040	5925	9399	5818	9348	5793
5242	5967	$1028\cdot 10^{-4}$	5858	$1022\cdot 10^{-4}$	5833
5451	6008	1115	5899	1109	5873
5669	6049	1203	5939	1196	5913
5896	6090	1289	5979	1282	5953
6131	6130	1376	6018	1368	5992
6376	6171	1461	6058	1453	6032
6631	6211	1547	6097	1538	6071
6897	6251	1632	6136	1623	6110
7172	6290	1716	6175	1707	6148
7459	6330	1800	6214	1790	6187
7757	6369	1884	6252	1874	6225
8067	6408	1967	6291	1956	6263
8390	6447	2050	6329	2039	6301
8725	6486	2132	6367	2121	6339
9074	6525	2214	6405	2202	6376
9437	6563	2296	6442	2283	6414
9814	6601	2377	6479	2364	6451
$1021\cdot 10^{-6}$	6639	2457	6517	2444	6488
1061	6677	2538	6554	2523	6525

$\rho_a DD$	$=0.01$		$=0.1$		$=0.5$
$\frac{\mu D}{a_s}$	$l=8$	$=12$	$=8$	$=12$	$=8$
$1104\cdot 10^{-6}$	$2768\cdot 10^{-4}$	$6855\cdot 10^{-4}$	$2752\cdot 10^{-4}$	$6830\cdot 10^{-4}$	$2684\cdot 10^{-4}$
1148	2852	6893	2836	6868	2766
1194	2936	6931	2919	6906	2847
1242	3019	6969	3002	6944	2927
1291	3102	7007	3084	6981	3007
1343	3184	7044	3166	7018	3087
1397	3266	7081	3247	7055	3166
1452	3347	7118	3328	7092	3245
1510	3428	7154	3409	7129	3324
1571	3509	7191	3489	7165	3401
1634	3589	7227	3568	7201	3479
1699	3668	7263	3647	7237	3556
1767	3747	7299	3726	7273	3633
1838	3826	7335	3804	7308	3709
1911	3904	7370	3882	7344	3785
1987	3982	7405	3959	7379	3860
2067	4059	7440	4036	7414	3935
2150	4136	7475	4113	7448	4009
2235	4212	7510	4189	7483	4083
2325	4288	7544	4264	7517	4157
2418	4364	7578	4339	7551	4230
2514	4439	7612	4414	7585	4303
2615	4514	7646	4488	7619	4375
2719	4588	7680	4562	7655	4447
2828	4662	7713	4635	7686	4518
2941	4735	7746	4708	7719	4589
3059	4808	7779	4781	7752	4660
3181	4880	7812	4853	7784	4730
3308	4952	7844	4924	7817	4799
3441	5023	7877	4995	7849	4869

Table IV (cont.)

$\rho_a DD =$	=0.5	=0.9		=0.99		$\rho_a DD =$	=0.01		=0.1		=0.5
$\frac{\mu D}{a_s}$	$l=12$	=8	=12	=8	=12	$\frac{\mu D}{a_s}$	$l=8$	=12	=8	=12	=8
$1104 \cdot 10^{-6}$	$6715 \cdot 10^{-4}$	$2617 \cdot 10^{-4}$	$6590 \cdot 10^{-4}$	$2603 \cdot 10^{-4}$	$6561 \cdot 10^{-4}$	$3578 \cdot 10^{-6}$	$5094 \cdot 10^{-4}$	$7909 \cdot 10^{-4}$	$5066 \cdot 10^{-4}$	$7881 \cdot 10^{-4}$	$4937 \cdot 10^{-4}$
1148	6752	2697	6627	2681	6597	3721	5165	7941	5136	7913	5006
1194	6789	2776	6663	2760	6634	3870	5235	7973	5206	7945	5074
1242	6826	2854	6699	2838	6670	4025	5305	8004	5275	7976	5141
1291	6863	2932	6735	2915	6705	4186	5374	8035	5344	8007	5208
1343	6899	3010	6771	2992	6741	4353	5443	8066	5412	8038	5275
1397	6936	3087	6807	3069	6776	4527	5511	8097	5480	8069	5341
1452	6972	3163	6842	3145	6811	4708	5579	8128	5548	8100	5407
1510	7008	3240	6877	3221	6846	4896	5647	8159	5615	8130	5472
1571	7044	3316	6912	3296	6881	5092	5714	8189	5681	8160	5537
1634	7079	3391	6947	3371	6916	5295	5780	8219	5748	8190	5601
1699	7115	3466	6982	3446	6950	5507	5846	8249	5813	8220	5665
1767	7150	3540	7016	3520	6984	5727	5912	8278	5879	8250	5729
1838	7185	3615	7050	3594	7018	5926	5977	8308	5943	8279	5792
1911	7220	3638	7084	3667	7052	6194	6042	8337	6008	8308	5855
1987	7254	3762	7118	3740	7086	6442	6106	8366	6072	8337	5917
2067	7289	3834	7152	3812	7119	6699	6170	8395	6135	8366	5979
2150	7323	3907	7185	3884	7152	6967	6233	8423	6198	8395	6040
2235	7357	3979	7218	3855	7185	7246	6296	8452	6261	8423	6101
2325	7391	4050	7251	4026	7218	7536	6359	8480	6323	8451	6162
2418	7424	4121	7284	4097	7251	7837	6421	8508	6385	8479	6222
2514	7457	4192	7317	4167	7283	8150	6482	8536	6446	8507	6281
2615	7491	4262	7349	4237	7315	8476	6543	8563	6507	8534	6341
2719	7524	4332	7381	4306	7347	8815	6604	8591	6567	8562	6400
2828	7556	4401	7413	4375	7379	9167	6664	8618	6627	8589	6458
2941	7589	4470	7445	4443	7411	9534	6724	8645	6687	8616	6516
3059	7621	4539	7477	4511	7442	9915	6783	8671	6746	8642	6573
3181	7654	4607	7508	4579	7474	$1031 \cdot 10^{-5}$	6842	8698	6804	8669	6630
3308	7686	4674	7540	4646	7505	1072	6901	8724	6862	8695	6687
3441	7717	4742	7571	4713	7536	1115	6959	8750	6920	8721	6743

Table IV (cont.)

$\rho_a DD$	=0.5	=0.9		=0.99		$\rho_a DD$	=0.01			=0.1	
$\frac{\mu D}{a_s}$	l=12	=8	=12	=8	=12	$\frac{\mu D}{a_s}$	l=4	=8	=12	=4	=8
$3578 \cdot 10^{-6}$	$7749 \cdot 10^{-4}$	$4808 \cdot 10^{-4}$	$7602 \cdot 10^{-4}$	$4779 \cdot 10^{-4}$	$7566 \cdot 10^{-4}$	$1160 \cdot 10^{-5}$	—	$7016 \cdot 10^{-4}$	$8776 \cdot 10^{-4}$	—	$6977 \cdot 10^{-4}$
3721	7780	4875	7632	4845	7597	1206	—	7073	8802	—	7034
3870	7812	4941	7663	4911	7627	1254	—	7130	8827	—	7091
4025	7843	5006	7693	4976	7657	1304	—	7186	8853	—	7147
4186	7873	5071	7723	5040	7687	1357	—	7242	8878	—	7202
4353	7904	5136	7753	5104	7717	1411	—	7297	8903	—	7257
4527	7934	5200	7783	5168	7746	1467	—	7352	8927	—	7312
4708	7965	5264	7812	5231	7776	1526	—	7407	8952	—	7366
4896	7995	5327	7842	5294	7805	1587	—	7460	8976	—	7419
5092	8024	5390	7871	5357	7834	1650	—	7514	9000	—	7473
5295	8054	5453	7900	5419	7862	1716	—	7567	9024	—	7525
5507	8084	5515	7929	5480	7891	1785	—	7620	9047	—	7578
5727	8115	5576	7957	5542	7919	1856	—	7672	9071	—	7630
5956	8142	5637	7985	5602	7947	1931	—	7723	9094	—	7681
6194	8171	5698	8014	5663	7975	2008	—	7775	9117	—	7732
6442	8199	5759	8042	5722	8003	2088	—	7825	9140	—	7783
6699	8228	5818	8069	5782	8030	2172	$8345 \cdot 10^{-5}$	7876	9162	$8252 \cdot 10^{-5}$	7833
6967	8256	5878	8097	5841	8058	2258	$1022 \cdot 10^{-4}$	7926	9184	$1011 \cdot 10^{-4}$	7883
7246	8284	5937	8124	5900	8085	2349	1208	7975	9207	1195	7932
7536	8312	5996	8151	5958	8112	2443	1392	8024	9228	1377	7981
7837	8339	6054	8178	6016	8139	2540	1575	8073	9250	1557	8029
8150	8367	6112	8205	6073	8166	2642	1756	8121	9272	1736	8077
8476	8394	6169	8232	6130	8192	2747	1935	8168	9293	1913	8124
8815	8421	6226	8258	6186	8218	2857	2113	8216	9314	2089	8171
9167	8448	6283	8284	6242	8244	2971	2289	8262	9335	2263	8218
9534	8474	6339	8310	6298	8270	3090	2463	8309	9355	2435	8264
9915	8501	6394	8336	6353	8296	3214	2636	8354	9376	2606	8310
$1031 \cdot 10^{-5}$	8527	6449	8362	6408	8321	3342	2807	8400	9396	2775	8355
1072	8553	6504	8387	6462	8346	3476	2976	8445	9416	2942	8400
1115	8579	6559	8413	6516	8371	3615	3143	8489	9436	3108	8444

Table IV (cont.)

$\rho_a DD=$	$=0.1$	$=0.5$			$=0.9$			$=0.99$		
$\frac{\mu D}{a_s}$	$l=12$	$=4$	$=8$	$=12$	$=4$	$=8$	$=12$	$=4$	$=8$	$=12$
$1160 \cdot 10^{-5}$	$8747 \cdot 10^{-4}$	—	$6799 \cdot 10^{-4}$	$8605 \cdot 10^{-4}$	—	$6613 \cdot 10^{-4}$	$8438 \cdot 10^{-4}$	—	$6570 \cdot 10^{-4}$	$8396 \cdot 10^{-4}$
1206	8773	—	6854	8630	—	6666	8462	—	6623	8421
1254	8798	—	6909	8655	—	6719	8487	—	6675	8445
1304	8824	—	6964	8680	—	6774	8512	—	6728	8470
1357	8849	—	7018	8705	—	6824	8536	—	6779	8494
1411	8873	—	7071	8730	—	6876	8560	—	6831	8518
1467	8898	—	7123	8754	—	6928	8584	—	6882	8541
1526	8922	—	7177	8779	—	6979	8608	—	6932	8565
1587	8947	—	7230	8803	—	7029	8631	—	6983	8588
1650	8971	—	7282	8826	—	7079	8654	—	7032	8611
1716	8994	—	7333	8850	—	7129	8677	—	7082	8634
1785	9018	—	7384	8874	—	7178	8700	—	7130	8657
1856	9041	—	7435	8897	—	7227	8723	—	7178	8679
1931	9065	—	7485	8920	—	7276	8746	—	7227	8702
2008	9088	—	7535	8943	—	7324	8768	—	7274	8724
2088	9110	—	7584	8965	—	7371	8790	—	7322	8746
2172	9133	$7860 \cdot 10^{-5}$	7633	8988	$7502 \cdot 10^{-5}$	7418	8812	$7426 \cdot 10^{-5}$	7368	8768
2258	9155	9626	7681	9010	9186	7465	8834	9092	7415	8789
2349	9177	$1138 \cdot 10^{-4}$	7729	9032	$1085 \cdot 10^{-4}$	7512	8855	$1074 \cdot 10^{-4}$	7461	8811
2443	9199	1311	7777	9054	1251	7558	8877	1238	7506	8832
2540	9221	1483	7824	9076	1414	7603	8898	1400	7551	8853
2642	9243	1653	7871	9097	1576	7648	8919	1560	7596	8874
2747	9264	1821	7917	9119	1737	7693	8940	1719	7640	8894
2857	9285	1988	7963	9140	1896	7737	8961	1876	7684	8915
2971	9306	2154	8009	9161	2053	7781	8981	2032	7727	8935
3090	9327	2317	8054	9181	2209	7824	9001	2186	7770	8955
3214	9347	2480	8098	9202	2363	7867	9021	2338	7813	8975
3342	9367	2640	8142	9222	2516	7910	9041	2489	7855	8994
3476	9387	2799	8186	9242	2667	7952	9061	2638	7897	9014
3615	9407	2956	8229	9262	2816	7994	9080	2786	7938	9033

Table IV (cont.)

$\rho_a DD=$	=0.01			=0.1			=0.5			=0.9	
$\frac{\mu D}{a_0}$	$l=4$	=8	=12	=4	=8	=12	=4	=8	=12	=4	=8
$3759 \cdot 10^{-5}$	$3309 \cdot 10^{-4}$	$8533 \cdot 10^{-4}$	$9455 \cdot 10^{-4}$	$3272 \cdot 10^{-4}$	$8488 \cdot 10^{-4}$	$9427 \cdot 10^{-4}$	$3112 \cdot 10^{-4}$	$8272 \cdot 10^{-4}$	$9282 \cdot 10^{-4}$	$2964 \cdot 10^{-4}$	$8035 \cdot 10^{-4}$
3910	3474	8577	9475	3434	8531	9446	3266	8315	9301	3110	8076
4066	3636	8620	9494	3595	8574	9466	3419	8357	9320	3255	8117
4228	3797	8663	9513	3754	8617	9485	3569	8398	9340	3398	8157
4398	3956	8705	9532	3911	8659	9503	3719	8440	9358	3539	8196
4573	4114	8747	9550	4067	8700	9522	3867	8480	9377	3679	8236
4756	4270	8788	9569	4221	8742	9540	4013	8521	9396	3818	8275
4946	4424	8829	9587	4374	8782	9559	4157	8561	9414	3955	8313
5144	4577	8869	9605	4525	8823	9577	4300	8600	9432	4090	8351
5350	4728	8909	9622	4674	8863	9594	4442	8639	9450	4224	8389
5564	4878	8949	9640	4822	8902	9612	4581	8678	9468	4356	8426
5786	5025	8988	9657	4968	8941	9629	4720	8716	9485	4487	8463
6017	5171	9026	9674	5112	8979	9646	4856	8754	9503	4616	8499
6258	5316	9065	9691	5255	9018	9663	4991	8791	9520	4744	8535
6508	5459	9102	9708	5396	9055	9680	5125	8828	9537	4870	8570
6768	5600	9140	9724	5535	9093	9697	5257	8865	9553	4994	8606
7039	5739	9177	9740	5673	9120	9713	5387	8901	9570	5117	8640
7320	5877	9213	9756	5800	9166	9729	5516	8936	9586	5239	8675
7613	6014	9249	9772	5944	9202	9745	5643	8972	9602	5359	8709
7917	6148	9284	9788	6077	9237	9761	5769	9007	9618	5477	8742
8233	6281	9319	9803	6208	9272	9776	5893	9041	9634	5594	8775
8563	6413	9354	9818	6338	9307	9791	6016	9075	9650	5709	8808
8905	6542	9388	9833	6466	9341	9806	6137	9109	9665	5823	8840
9261	6671	9422	9848	6593	9374	9821	6257	9142	9680	5936	8872
9632	6797	9455	9862	6718	9408	9836	6375	9174	9695	6046	8903
$1002 \cdot 10^{-4}$	6922	9488	9877	6841	9440	9850	6491	9207	9710	6156	8934
1042	7045	9520	9891	6963	9473	9865	6606	9239	9724	6264	8965
1083	7167	9552	9905	7083	9505	9879	6719	9270	9739	6370	8995
1127	7287	9584	9919	7202	9536	9892	6831	9301	9753	6475	9025
1172	7405	9615	9932	7319	9667	9906	6941	9332	9767	6578	9055

Table IV (cont.)

$\epsilon_a DD=$	$=0.9$	$=0.99$			$\epsilon_u DD=$	$=0.9$			$=0.99$		
$\frac{\mu D}{a_z}$	$l=12$	$=4$	$=8$	$=12$	$\frac{\mu D}{a_z}$	$l=4$	$=8$	$=12$	$=4$	$=8$	$=12$
$3759 \cdot 10^{-5}$	$9099 \cdot 10^{-4}$	$2932 \cdot 10^{-5}$	$7979 \cdot 10^{-4}$	$9052 \cdot 10^{-4}$	$3950 \cdot 10^{-4}$	$9076 \cdot 10^{-4}$	$9774 \cdot 10^{-4}$	$9914 \cdot 10^{-4}$	$8952 \cdot 10^{-4}$	$9695 \cdot 10^{-4}$	$9857 \cdot 10^{-4}$
3910	9119	3077	8020	9071	4108	9135	9792	9923	9010	9712	9865
4066	9137	3220	8060	9090	4272	9193	9809	9931	9067	9728	9873
4228	9156	3361	8099	9108	4443	9250	9825	9938	9122	9744	9881
4398	9175	3501	8138	9127	4620	9305	9842	9946	9176	9760	9889
4537	9193	3639	8177	9145	4805	9359	9857	9954	9229	9776	9896
4756	9211	3776	8216	9163	4997	9412	9873	9961	9281	9791	9903
4946	9229	3911	8254	9181	5197	9464	9888	9968	9331	9806	9910
5144	9247	4045	8291	9198	5405	9515	9903	9975	9381	9821	9917
5350	9264	4177	8329	9216	5621	9564	9918	9982	9429	9835	9924
5564	9282	4308	8365	9233	5845	9612	9932	9989	9476	9849	9931
5786	9299	4437	8402	9250	6079	9660	9946	9995	9522	9863	9937
6017	9316	4564	8438	9267	6322	9706	9960	10000	9567	9876	9943
6258	9333	4690	8473	9283	6575	9751	9973	10000	9611	9889	9950
6508	9349	4814	8508	9300	6838	9795	9986	10000	9654	9902	9956
6768	9366	4937	8543	9316	7111	9837	9999	10000	9696	9914	9962
7039	9382	5059	8577	9332	7395	9879	10000	10000	9736	9926	9967
7320	9398	5179	8611	9348	7691	9920	10000	10000	9776	9938	9973
7613	9414	5297	8645	9364	7999	9960	10000	10000	9815	9950	9978
7917	9430	5414	8678	9379	8318	9998	10000	10000	9853	9961	9984
8233	9445	5529	8710	9394	8651	10000	10000	10000	9889	9972	9989
8563	9460	5643	8743	9409	8997	10000	10000	10000	9925	9982	9994
8905	9475	5755	8775	9424	9356	10000	10000	10000	9960	9993	9999
9261	9490	5866	8806	9439	9730	10000	10000	10000	9994	10000	10000
9632	9505	5975	8837	9454							
$1002 \cdot 10^{-4}$	9520	6083	8868	9468							
1042	9534	6189	8898	9482							
1083	9548	6294	8928	9496							
1127	9562	6397	8957	9510							
1172	9576	6499	8986	9524							

Table V

t	$(1-\rho_b DD) = 10^{-4}$	$=10^{-3}$	$=2\cdot10^{-3}$	$=3\cdot10^{-3}$	$=4\cdot10^{-3}$	$=5\cdot10^{-3}$
0.1	$9524\cdot10^{-4}$	$9533\cdot10^{-4}$	$9523\cdot10^{-4}$	$9522\cdot10^{-4}$	$9522\cdot10^{-4}$	$9521\cdot10^{-4}$
0.2	9091	9090	9089	9088	9087	9086
0.3	8696	8694	8693	8691	8690	8688
0.4	8333	8331	8329	8327	8325	8323
0.5	8000	7998	7995	7993	7990	7988
0.6	7692	7689	7686	7683	7681	7678
0.7	7407	7404	7401	7397	7394	7390
0.8	7142	7139	7135	7131	7127	7123
0.9	6896	6892	6888	6884	6879	6875
1.0	6666	6662	6657	6652	6647	6643
1.1	6451	6446	6441	6436	6431	6425
1.2	6249	6244	6239	6233	6227	6222
1.3	6060	6054	6048	6042	6036	6030
1.4	5882	5876	5869	5863	5856	5849
1.5	5714	5707	5700	5693	5686	5679
1.6	5555	5548	5541	5533	5526	5518
1.7	5405	5398	5390	5382	5374	5366
1.8	5262	5255	5247	5238	5230	5222
1.9	5127	5119	5111	5102	5093	5085
2.0	4999	4991	4982	4973	4964	4954
2.2	4761	4752	4742	4732	4722	4712
2.4	4544	4535	4524	4513	4503	4492
2.6	4347	4336	4325	4313	4302	4290
2.8	4165	4154	4142	4130	4117	4105
3.0	3997	3987	3974	3961	3948	3935

Table V (cont.)

l	$(1-\rho_b DD) = 10^{-4}$	$=10^{-3}$	$=2\cdot10^{-3}$	$=3\cdot10^{-3}$	$=4\cdot10^{-3}$	$=5\cdot10^{-3}$
3.2	$3845\cdot10^{-4}$	$3832\cdot10^{-4}$	$3818\cdot10^{-4}$	$3805\cdot10^{-4}$	$3791\cdot10^{-4}$	$3777\cdot10^{-4}$
3.4	3702	3689	3674	3660	3645	3631
3.6	3570	3556	3541	3525	3510	3495
3.8	3447	3432	3416	3400	3384	3368
4.0	3332	3316	3300	3283	3266	3250
5.0	2855	2837	2816	2796	2776	2756
6.0	2498	2476	2452	2428	2405	2382
7.0	2219	2195	2167	2141	2114	2089
8.0	1997	1969	1938	1909	1880	1851
9.0	1815	1784	1750	1717	1686	1655
10	1663	1629	1592	1557	1522	1489
12	1424	1384	1342	1301	1262	1225
14	1245	1199	1152	1106	1064	1023
16	1105	1054	1002	$9523\cdot10^{-5}$	$9065\cdot10^{-5}$	$8636\cdot10^{-5}$
18	$9935\cdot10^{-5}$	$9372\cdot10^{-5}$	$8798\cdot10^{-5}$	8272	7789	7343
20	9019	8475	7789	7235	6733	6278
30	6145	5303	4544	3926	3416	2990

Table V (cont.)

l	$(1-\rho_b DD)=6\cdot10^{-3}$	$=7\cdot10^{-3}$	$=8\cdot10^{-3}$	$=9\cdot10^{-3}$	$=1\cdot10^{-2}$	$=2\cdot10^{-2}$	$=3\cdot10^{-2}$	$=4\cdot10^{-2}$
0.1	$9521\cdot10^{-4}$	$9520\cdot10^{-4}$	$9520\cdot10^{-4}$	$9519\cdot10^{-4}$	$9519\cdot10^{-4}$	$9514\cdot10^{-4}$	$9509\cdot10^{-4}$	$9504\cdot10^{-4}$
0.2	9085	9084	9083	9082	9081	9071	9061	9051
0.3	8687	8685	8683	8682	8681	8666	8651	8636
0.4	8321	8319	8318	8316	8314	8294	8274	8255
0.5	7985	7983	7980	7978	7975	7951	7927	7902
0.6	7675	7672	7669	7666	7663	7634	7605	7576
0.7	7387	7384	7380	7377	7373	7340	7306	7273
0.8	7120	7116	7112	7108	7104	7066	7028	6990
0.9	6871	6866	6862	6858	6853	6810	6768	6726
1.0	6638	6633	6628	6624	6619	6572	6525	6479
1.1	6420	6415	6410	6404	6399	6348	6297	6246
1.2	6216	6210	6205	6199	6193	6137	6082	6028
1.3	6024	6018	6012	6066	5999	5939	5880	5822
1.4	5843	5836	5830	5823	5817	5753	5689	5627
1.5	5672	5665	5658	5651	5645	5576	5509	5443
1.6	5511	5504	5496	5489	5482	5409	5338	5269
1.7	5358	5351	5343	5335	5327	5250	5176	5103
1.8	5214	5205	5197	5189	5181	5101	5022	4946
1.9	5076	5068	5059	5050	5042	4958	4876	4796
2.0	4945	4936	4927	4919	4910	4822	4736	4653
2.2	4703	4693	4683	4673	4664	4568	4476	4387
2.4	4481	4471	4460	4450	4439	4337	4238	4143
2.6	4279	4268	4256	4245	4234	4125	4020	3918
2.8	4093	4081	4069	4057	4045	3929	3818	3712
3.0	3922	3909	3896	3884	3871	3749	3632	3520

Table V (cont.)

l	$(1-\rho_b DD)=6\cdot10^{-3}$	$=7\cdot10^{-3}$	$=8\cdot10^{-3}$	$=9\cdot10^{-3}$	$=1\cdot10^{-2}$	$=2\cdot10^{-2}$	$=3\cdot10^{-2}$	$=4\cdot10^{-2}$
3.2	$3764\cdot10^{-4}$	$3750\cdot10^{-4}$	$3737\cdot10^{-4}$	$3723\cdot10^{-4}$	$3710\cdot10^{-4}$	$3581\cdot10^{-4}$	$3459\cdot10^{-4}$	$3342\cdot10^{-4}$
3.4	3617	3603	3588	3574	3560	3425	3298	3177
3.6	3480	3465	3450	3436	3421	3280	3147	3022
3.8	3353	3337	3322	3306	3291	3144	3007	2878
4.0	3233	3217	3201	3185	3169	3017	2875	2742
5	2737	2717	2698	2679	2660	2482	2321	2175
6	2360	2337	2315	2294	2272	2073	1898	1744
7	2063	2038	2014	1990	1966	1750	1566	1408
8	1823	1796	1769	1743	1718	1488	1300	1142
9	1625	1595	1566	1539	1511	1272	1083	$9296\cdot10^{-5}$
10	1457	1426	1395	1366	1337	1092	$9046\cdot10^{-5}$	7580
12	1189	1155	1122	1090	1060	$8109\cdot10^{-5}$	6347	5058
14	$9849\cdot10^{-5}$	$9485\cdot10^{-5}$	$9139\cdot10^{-5}$	$8811\cdot10^{-5}$	$8498\cdot10^{-5}$	6061	4472	3384
16	8235	7858	7505	7174	6861	4546	3156	2266
18	6932	6551	6199	5871	5566	3417	2230	1518
20	5863	5485	5139	4821	4529	2571	1576	1018
30	2631	2327	2067	1843	1649	$6240\cdot10^{-6}$	$2788\cdot10^{-6}$	$1377\cdot10^{-6}$

APPENDIX

The following text may be read independently of the rest of the book. The results of the preceding sections constitute examples of simple and special relations, illustrating the more general laws enunciated in this appendix.

Statement of the Problem

We are interested in the physical essence of the flow of particles contained in a vessel. The vessel may have any shape and dimensions desired, and it may be closed or connected to a certain number of reservoirs, which serve as sources of new reacting particles or sinks for reaction products.

The physicochemical characteristics of different portions of the surface of the reaction vessel and the contents of the reservoirs, generally speaking, can vary with the passage of time and can differ at different points in space. The walls of the vessel may also be regarded as particle sources or sinks. The various portions of the surface can act as catalysts for certain types of processes.

Before undertaking our investigation in so general a framework, let us consider the problem in a simpler vein, namely the motion of particles of a single type which are incapable of chemical conversion but move in vessels of complex configuration.

Molecular Flows of Monotypical Particles

It is useful to assign a classification (Polyak, 1935; Surinov, 1950*) to the possible types of molecular flux per square centimeter of the i-th area. The classification has already been utilized in Sections 1,4, and elsewhere, but it was not explicitly formulated.

1. Natural (Langmuir) emission, vaporization or desorption E_{ci}, determined by the temperature of the emitting surface:

$$E_{ci} = \alpha_i E_{0i} , \tag{A.1}$$

where α_i is the condensation coefficient, E_{0i} is the equilibrium emission.

2. Absorbed (or condensed) molecular flux:

$$A_{ni} = I_i(1 - \rho_i) = I_i \alpha_i , \tag{A.2}$$

where I_i is the incident molecular flux, $\rho_i = 1 - \alpha_i$ is the reflection coefficient.

3. Reflected molecular flux:

$$R_i = I_i \rho_i . \tag{A.3}$$

4. Effective molecular flux:

$$E_e = E_{ci} + R_i = \alpha E_{0i} + I_i \rho_i . \tag{A.4}$$

*The works of Polyak, Surinov, and certain other authors cited in the Appendix relate to photometry. We have translated them into the terminology of molecular flow theory.

5. Incident molecular flux on the i-th surface area:

$$I_i = \sum_{k=1}^{r} E_{ek}\,\varphi_{ik} \quad (k = 1,2\ldots r)\,, \tag{A.5}$$

where E_{ek} is the effective molecular flux arriving from the k-th area (φ_{ik} are proportionality factors, which will be defined in the next section).

6. Resultant absorption:

$$A_{pi} = A_{ni} - E_{ci} = \alpha(I_i - E_{oi})\,. \tag{A.6}$$

We also give a few simple corollaries of the foregoing relations:

$$A_{pi} = I_i - E_{ei}\;; \tag{A.6a}$$

$$A_{ni} = A_{pi} + E_{ci}\;; \tag{A.6b}$$

$$I_i = \frac{A_{pi}}{\alpha_i} + E_{oi}\;; \tag{A.7}$$

$$E_{ei} = \frac{\rho_i A_{pi}}{\alpha_i} + E_{oi}\,. \tag{A.8}$$

Some Geometric (Photometric) Relations

Equation (A5) involves the coefficient φ_{ik}, the so-called coupling coefficient or angular coefficient, indicating what fraction of the molecular flux (e.g., E_{ei}) arriving from an area of type i is incident on the area of type k. The magnitude of the coupling coefficient depends on the size and relative orientation of the i-th and k-th areas, as well as on the particle emission and reflection characteristics.

The magnitude and form of the characteristics depend on their method of normalization. In Eq. (1.4) the characteristic is equal to the ratio of the flux in a given direction to the flux normal to the surface, given the law of cosines. It could be defined as the ratio of the molecular flux in a given direction to the total flux E arriving per unit surface. We will give the dependence of the intensity of a molecular beam on direction, letting

$$\bar{E}(\theta, \gamma, \bar{x}) = \bar{E}(0, \gamma, \bar{x}) f_1(\theta, \gamma, \bar{x})\,, \tag{A.9}$$

where $\bar{E}$ is taken per unit surface area per unit solid angle.

Then, with the second method of normalization, the characteristic is determined from the relation

$$f_2(\theta, \gamma, \bar{x}) = \frac{\bar{E}(\theta, \gamma, \bar{x})}{E} = \frac{\bar{E}(\theta, \gamma, \bar{x})}{\bar{E}(0, \gamma, \bar{x}) \int_{\Omega} f_1 d\Omega}\,.$$

If we denote

$$\bar{\Omega} = \int_{\Omega} f_1 d\Omega\,, \tag{A.10}$$

then

$$f_2 = \frac{f_1}{\bar{\Omega}} \,. \tag{A.11}$$

Other methods of normalization could also be used. We let $\bar{\Omega}$ signify the equivalent angle, giving the ratio of the total molecular flux to the flux normal to a given area.

If we assume that the characteristic does not vary from point to point, the angle $\bar{\Omega}$ will be determined according to the equation

$$\bar{\Omega} = \int_0^{k\pi} d\theta \int_0^{2\pi} f_1(\theta, \gamma) \sin\theta \, d\gamma \,, \tag{A.12}$$

where $k = {}^1/_2$ if the area radiates into one hemisphere and $k = 1$ if the area is completely surrounded by space.

For emission according to the cosine law*

$$f(\theta) = \cos\theta, \quad \bar{\Omega}_{k=1} = 2\pi, \quad \bar{\Omega}_{k=½} = \pi \,.$$

For emission according to the Euler law, $f(\theta) = 1$,

$$\bar{\Omega}_{k=½} = 2\pi; \quad \bar{\Omega}_{k=1} = 4\pi \,.$$

The angular coefficient is determined from the equation

$$\varphi_{ik} = \frac{1}{S_i} \int\limits_{(S_i)} \int\limits_{(S_k)} \frac{f(\theta_i, \gamma_i, \bar{x}_i) \cos\theta_k dS_i dS_k}{\bar{\Omega} l_{ik}^2} = \frac{1}{S_i} \int\limits_{(S_i)} \int\limits_{(S_k)} \frac{f(\theta_i, \gamma_i, \bar{x}_i) \, dS_i \, d\Omega_i}{\bar{\Omega}} \,. \tag{A.13}$$

The quantity

$$S_{ik} = \varphi_{ik} S_i \tag{A.14}$$

is called the <u>mutual surface</u> of the i-th and k-th bodies. Under equilibrium conditions the so-called <u>reciprocity principle</u> is valid:

$$S_{ik} = S_{ki} \,. \tag{A.15}$$

We now introduce the <u>probability of incidence</u> of a particle emitted according to the given law (A9) from the area dS_1 on the surface S_2:

$$P_{dS_1 - S_2} = \frac{1}{\bar{\Omega}} \int\limits_{\Omega} f(\theta, \gamma, \bar{x}_1) d\Omega \,. \tag{A.16}$$

---

* We will be concerned below only with f_1 and will write, simply, f.

In the special case of the law of cosines we obtain from Eq. (A16) the following relation, cited in Lozgachev, 1962 (4):

$$P_{dS_1 - S_2} = \frac{1}{2\pi}\int_{\gamma}^{\gamma_2} d\gamma \int_{\theta(\gamma)}^{\theta_2(\gamma)} \sin 2\theta\, d\theta\ , \tag{A.17}$$

where γ_1, γ_2 and θ_1, θ_2 are the limiting angles at which the surface S_2 can still be "seen" by dS_1.

Once the angular coefficients φ_{ik} are known, the molecular fluxes proceeding from one surface to another can be calculated for a specific system of bodies.

It is apparent from Eqs. (A.13) and (A.16) that

$$\varphi_{ik} = \frac{1}{S_i}\int_{(S_i)} P_{dS_i - S_k}\, dS_i\ . \tag{A.18}$$

Table A1 gives certain values of the coefficients ϕ_{ik} (according to Blokh, 1962) calculated by means of Knudsen's law.

If the system of surfaces i, k form a closed figure or if one of them, say the k-th, is emitting, we then have

$$\sum_{i=1}^{r} \varphi_{ik} = 1\ , \tag{A.19}$$

which is a consequence of the law of conservation of matter.

Making use of Eq. (A.17), Lozgachev [1962 (4)] derived the following equation for the special case of irradiation of a plane by an evaporator in the form of a disk of radius r, situated at a distance h from the plane in question:

$$P_{\rho}^{\mathrm{or}}(h) = \tfrac{1}{2}\left(1 - \frac{h^2 - r^2 + \rho^2}{\sqrt{(h^2 + r^2 + \rho^2)^2 - 4r^2\rho^2}}\right), \tag{A.20}$$

which states the probability of incidence of particles from the disk at a point on the plane situated at a distance ρ from the axis of the disk.

We will also give an additional result of Lozgachev from the same work (see Fig. A.1)

The probability of incidence of particles from the same disk at a point M situated in a plane perpendicular to the plane of the disk is expressed by the relation

$$P^{\mathrm{or}}(abl) = \frac{bl}{2(a^2 + b^2)}\left(\frac{a^2 + b^2 + l^2 + r^2}{\sqrt{(a^2 + b^2 + l^2 + r^2)^2 - 4r^2(a^2 + b^2)}} - 1\right)\ . \tag{A.21}$$

Systems of Algebraic Equations for Molecular Fluxes

We obtain the systems of equations corresponding to various types [see the text relating to Eqs. (A.1)-(A.2)] of molecular flow by combining Eqs. (A.1)-(A.8).

Table A1. Relations for Determining Coupling Coefficients

Entry No.	Relative orientation and shape of surfaces	Coupling coefficients and mutual surfaces
1	2	3
1.	Two surfaces forming a closed system, surface #1 having no concavities. 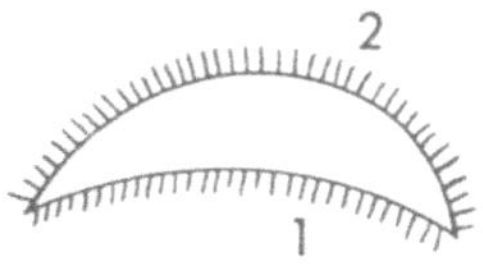	$\varphi_{12} = 1;\ \varphi_{21} = \frac{S_1}{S_2};$ $S_{12} = S_1.$
2.	A body having no concavities and enclosed within another body. 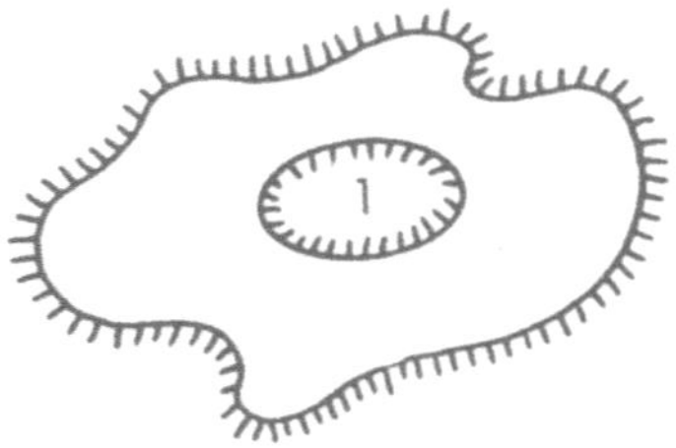	$\varphi_{12} = 1;\ \varphi_{21} = \frac{S_1}{S_2};$ $S_{12} = S_1.$
3.	Two arbitrarily oriented plane areas, separated by a distance large in comparison with their dimensions; normals to the centers of the areas in the same plane. 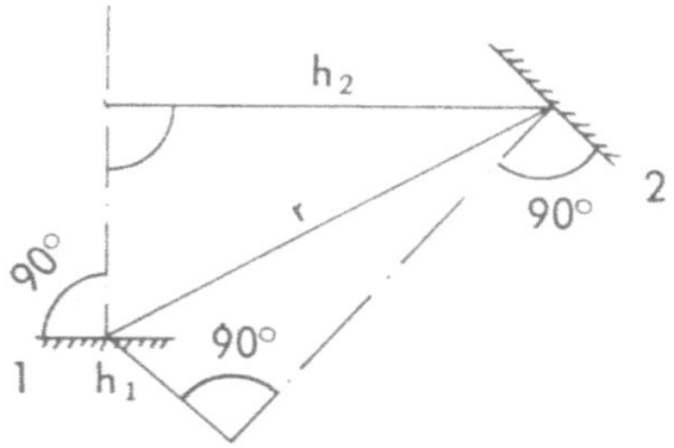	$\varphi_{12} = \frac{h_1 h_2}{\pi r^4} S_2;$ $S_{12} = \varphi_{12} S_1.$

Table A1 (continued)

Entry No.	Relative orientation and shape of surfaces	Coupling coefficients and mutual surfaces
1	2	3
4.	Two identical rectangles situated opposite one another in parallel planes. 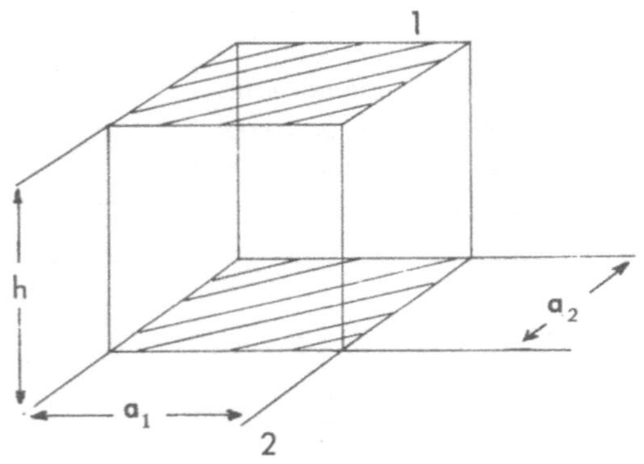	$\varphi_{12} = \varphi_{21} = f^1(A_1, A_2)$ $= \frac{2}{\pi}\left[\frac{\sqrt{1+A_1^2}}{A_1}\tan^{-1}\frac{A_2}{\sqrt{1+A_1^2}} + \frac{\sqrt{1+A_2^2}}{A_2}\tan^{-1}\frac{A_1}{\sqrt{1+A_2^2}} - \frac{1}{A_1}\tan^{-1}A_2 - \frac{1}{A_2}\tan^{-1}A_1 + \frac{1}{2A_1A_2}\ln\frac{(1+A_1^2)(1+A_2^2)}{1+A_1^2+A_2^2}\right],$ where $A_1 = \frac{a_1}{h}$; $A_2 = \frac{a_2}{h}$; $S_{12} = a_1 a_2 \varphi_{12}$. For squares: $a_1 = a_2 = a$; $A_1 = A_2 = A$; $\varphi_{12} = \frac{4}{\pi}\left[\frac{\sqrt{1+A^2}}{A}\tan^{-1}\frac{A}{\sqrt{1+A^2}} - \frac{1}{A}\tan^{-1}A + \frac{1}{4A^2}\ln\frac{(1+A^2)^2}{1+2A^2}\right],$ $S_{12} = a^2\varphi_{12}$.
5.	Two circular disks in parallel planes with centers on a common normal. 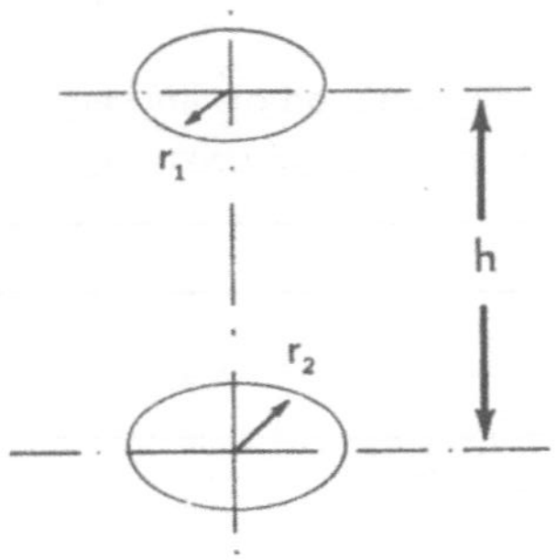	$\varphi_{12} = \frac{r_1^2 + r_2^2 + h^2 - \sqrt{(r_1^2 + r_2^2 + h^2)^2 - 4r_1^2 r_2^2}}{2r_1},$ $S_{12} = \frac{\pi h}{4}\left[\sqrt{h^2 + (r_1+r_2)^2} - \sqrt{(r_2-r_1)^2 + h^2}\right].$

Table A1 (continued)

Entry No.	Relative orientation and shape of surfaces	Coupling coefficients and mutual surfaces
1	2	3
6.	Two mutually perpendicular rectangles with a common side. 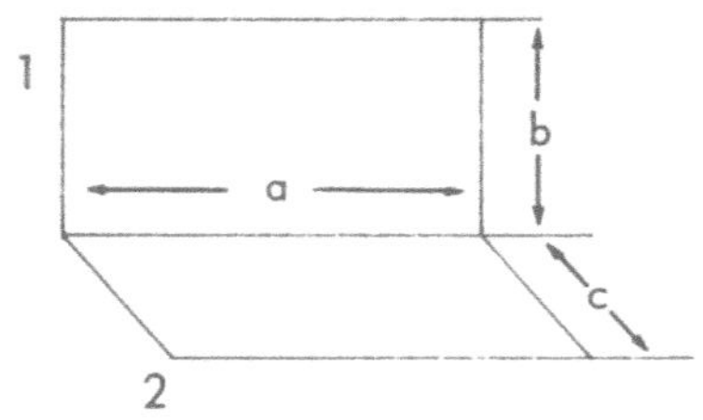	$\varphi_{12} = \frac{1}{\pi}\left[\tan^{-1}\frac{1}{B} + \frac{C}{B}\tan^{-1}\frac{1}{C} - \sqrt{C^2-1}\,\tan^{-1}\frac{1}{\sqrt{B^2+C^2}} + \frac{C^2}{4B}\ln\frac{C^2(1+B^2+C^2)}{(1+C^2)(B^2+C^2)} + \frac{B}{4}\ln\frac{B^2(1+B^2+C^2)}{(1+C^2)(B^2+C^2)} - \frac{1}{4}\ln\frac{1+B^2+C^2}{(1+B^2)(1+C^2)}\right],$ where $B = \frac{b}{a}$; $C = \frac{c}{a}$; $S_{12} = ab\varphi_{12}$.
7.	Elementary area dS and rectangle S in parallel planes; one corner of the rectangle on the normal to the center of the elementary area. 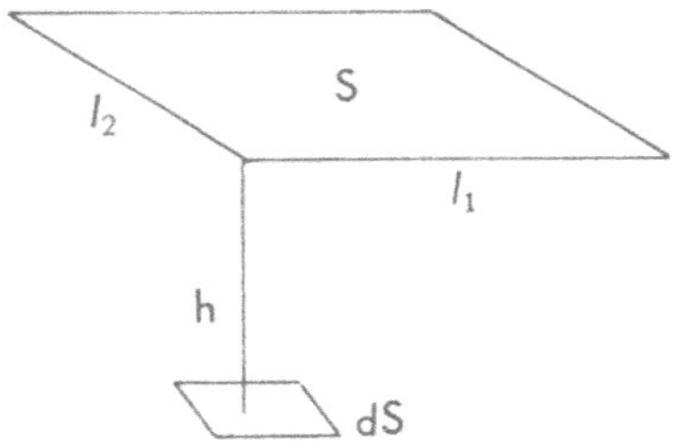	$\varphi_{dS\text{-}S} = \frac{1}{2\pi}\left[\alpha_1 \tan^{-1}\frac{\alpha_1}{L_1}L_2 + \alpha_2\tan^{-1}\frac{\alpha_2}{L_2}L_1\right],$ where $\frac{l_1}{h} = L_1$; $\frac{l_2}{h} = L_2$; $\alpha_1 = \frac{L_1}{\sqrt{1+L_1^2}}$; $\alpha_2 = \frac{L_2}{\sqrt{1+L_2^2}}$; $\varphi_{S\text{-}dS} = \varphi_{dS\text{-}S}\frac{dS}{S}$.

Table A1 (continued)

Entry No.	Relative orientation and shape of surfaces	Coupling coefficients and mutual surfaces
1	2	3
8.	Elementary area dS and rectangle S in perpendicular planes. 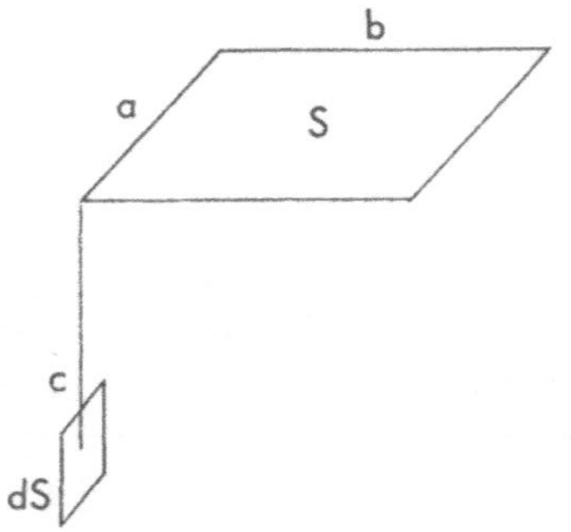	$\varphi_{dS\text{-}S} = \frac{1}{2\pi}\left[\operatorname{cosec}\frac{1}{\sqrt{1+C^2}} - \frac{1}{\sqrt{1+B^2+C^2}}\operatorname{cosec}\frac{1}{\sqrt{1+B^2+C^2}}\right]$ where $B = \frac{b}{a}$; $C = \frac{c}{a}$; for $B \to \infty$ $\varphi_{dS\text{-}S} = \frac{1}{2\pi}\operatorname{cosec}\frac{1}{\sqrt{1+C^2}}$; $S_{dS\text{-}S} = \varphi_{dS\text{-}S}\,dS$.
9.	Elementary area dS and circular disk S in parallel planes. 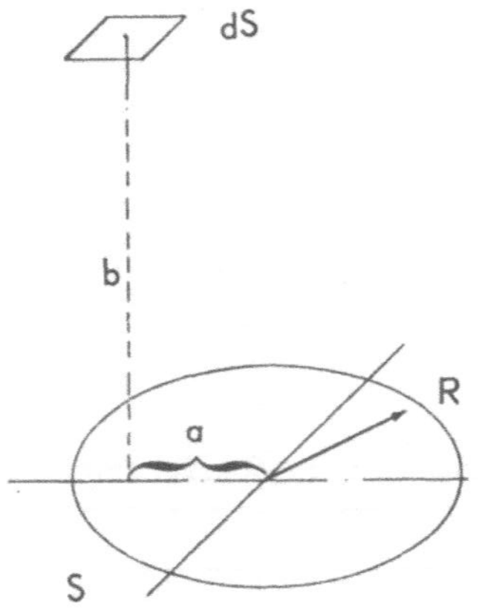	$\varphi_{dS\text{-}S} = \frac{1}{2}\left[1 - \frac{A^2 + B^2 - 1}{\sqrt{(1+A^2+B^2)^2 + 4A^2}}\right]$ where $A = \frac{a}{R}$; $B = \frac{b}{R}$; $S_{dS\text{-}S} = \varphi_{dS\text{-}S}\,dS$; For $a = 0$, $A = 0$, $\varphi_{dS\text{-}S} = \frac{1}{1+B^2}$.

Table A1 (continued)

Entry No.	Relative orientation and shape of surfaces	Coupling coefficients and mutual surfaces
1	2	3
10.	Elementary area dS and circular disk S in perpendicular planes. 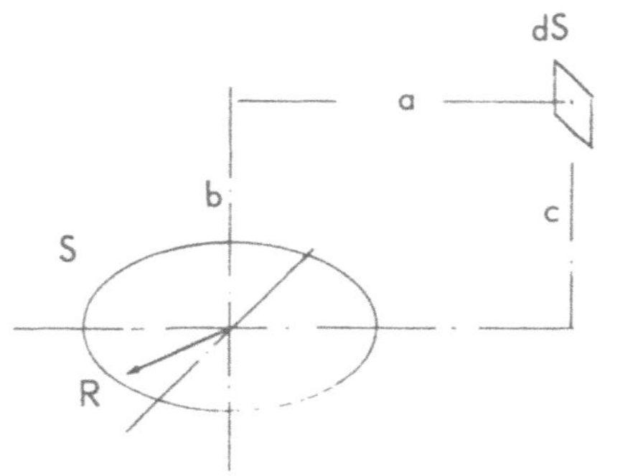	$\varphi_{dS-S} = \frac{C}{2A}\left[\frac{1 + A^2 + C^2}{\sqrt{(1 + A^2 + C^2)^2 + 4A^2}}\right]$ where $A = \frac{a}{R}$; $C = \frac{c}{R}$; $S_{dS-S} = \varphi_{dS-S}\, dS$.
11.	Two parallel cylinders of equal diameter d. 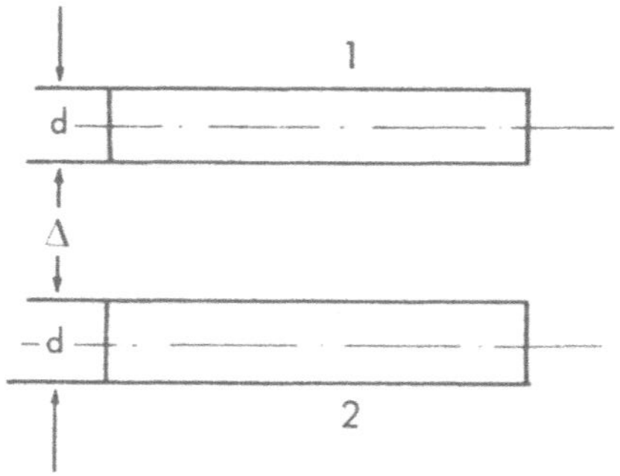	$\varphi_{12} = \varphi_{21} = \frac{1}{\pi}\left[\sin^{-1} D + \sqrt{\frac{1}{D^2} + 1} - \frac{1}{D}\right]$, where $D = \frac{d}{\Delta}$; $S_{12} = \Delta\left[\sqrt{1 - D^2} + D\left(\sin^{-1} D\right) - 1\right]$. The surface S_{12} is normalized to one meter length of the cylinders.

Table A1 (continued)

Entry No.	Relative orientation and shape of surfaces	Coupling coefficients and mutual surfaces
1	2	3
12.	Unbounded plane and a row of cylinders in a parallel plane. 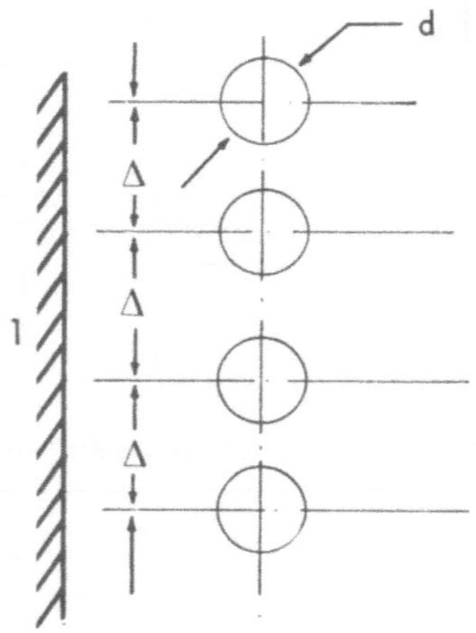 	$\varphi_{12} = 1 - \sqrt{1 - \left(\frac{d}{\Delta}\right)^2} + \frac{d}{\Delta} \tan^{-1} \sqrt{\left(\frac{\Delta}{d}\right)^2 - 1}$; $\varphi_{21} = \frac{1}{\pi} \left[\frac{\Delta}{d} - \sqrt{\left(\frac{\Delta}{d}\right)^2 - 1} + \tan^{-1} \sqrt{\left(\frac{\Delta}{d}\right)^2 - 1} \right]$; $S_{12} = S_{21} = \varphi_{12}\,\Delta = \varphi_{21}\, d$. The surface S_{12} is normalized to one tube and one meter length of tube.
13.	Unbounded plane and two rows of cylinders in parallel planes. 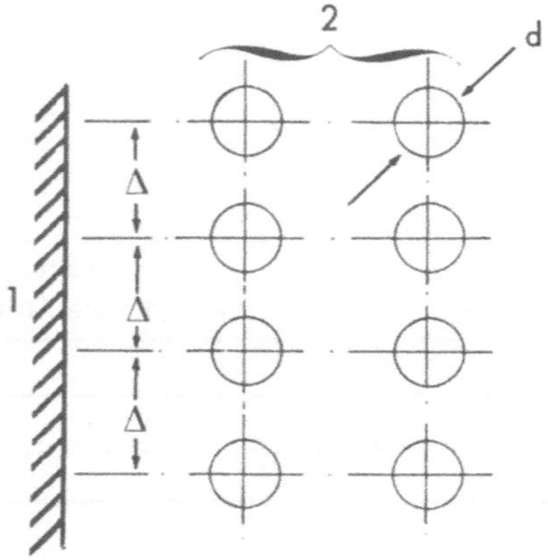 	$\varphi_{12} = 1 - (1 - \varphi'_{12})^2$, where φ'_{12} is the coupling coefficient for one row of cylinders (cf. Entry #12); $S_{12} = \varphi_{12}\,\Delta$; For n rows of cylinders $\varphi_{12} = 1 - (1 - \varphi'_{12})^n$.

Table A1 (continued)

Entry No.	Relative orientation and shape of surfaces	Coupling coefficients and mutual surfaces
1	2	3
14.	Convex body situated between two parallel planes; dimensions of the body small in comparison with dimensions of the planes. 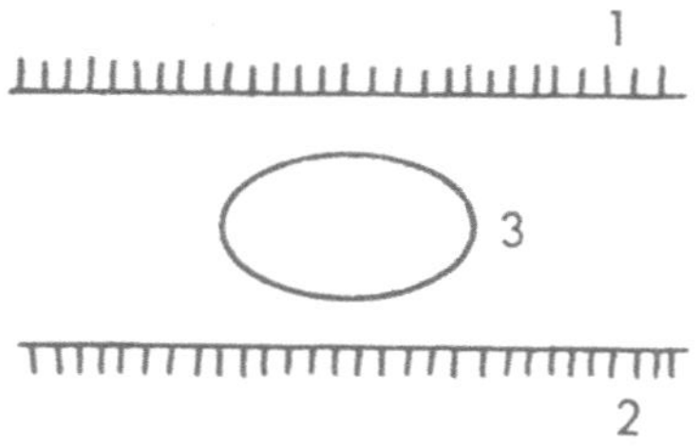	$\varphi_{12} = \varphi_{21} = 1$; $\varphi_{23} = \varphi_{13} = 0$; $S_{13} = S_{31} = S_{23} = S_{32} = \frac{1}{2} S_3$; $S_{12} = S_1 = S_2$.
15.	Three nonconcave surfaces forming a closed system of infinite extent. 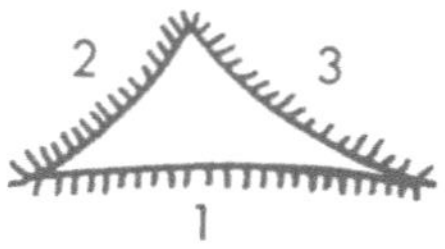	$\varphi_{12} = \left(1 + \frac{S_2}{S_1} + \frac{S_3}{S_2}\right)$; $S_{12} = \frac{1}{2}(S_1 + S_2 - S_3)$; $\varphi_{21} = \left(1 + \frac{S_1}{S_2} - \frac{S_3}{S_2}\right)$, $S_{21} = \frac{1}{2}(S_1 + S_2 - S_3)$; $\varphi_{13} = \left(1 + \frac{S_3}{S_1} - \frac{S_2}{S_1}\right)$; $S_{13} = \frac{1}{2}(S_1 + S_3 - S_2)$; $\varphi_{23} = \left(1 + \frac{S_3}{S_2} - \frac{S_1}{S_2}\right)$; $S_{23} = \frac{1}{2}(S_2 + S_3 - S_1)$.
16.	Four nonconcave surfaces forming a closed system of infinite extent. 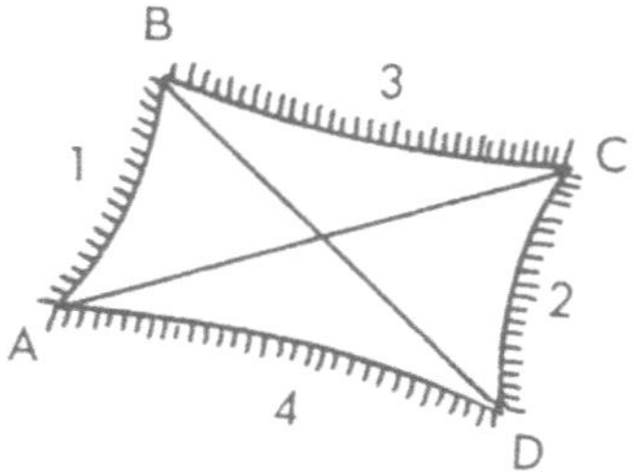	$\varphi_{kn} = \frac{S_{kn}}{S_k}$; $S_{12} = \frac{1}{2}(S_{AC} + S_{BD} - S_3 - S_4)$; $S_{13} = \frac{1}{2}(S_1 + S_3 - S_{AC})$; $S_{14} = \frac{1}{2}(S_1 + S_4 - S_{BD})$; etc.

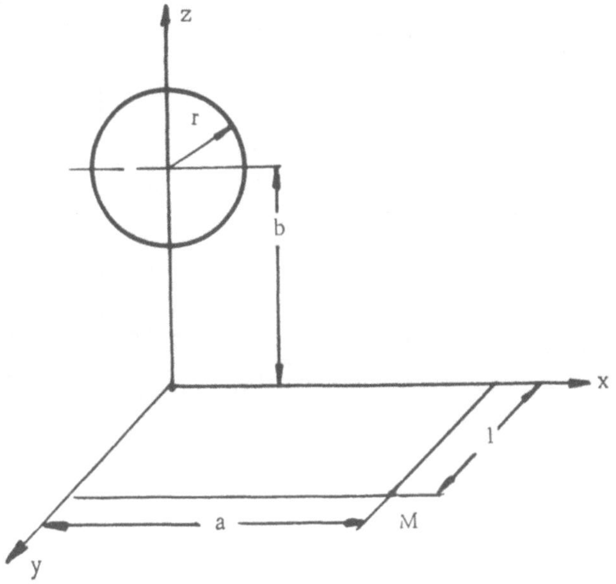

Fig. A.1. Position of the evaporator relative to the plane of the condenser (receiver) (Lozgachev).

a. System of equations for incident fluxes; into Eq. (A.5) we substitute the value of E_e; according to (A.4),

$$I_i - \sum_{k=1}^{r} \rho_k I_k \varphi_{ik} = \sum_{k=1}^{r} E_{ck} \varphi_{ik}. \tag{A.22}$$

Multiplying all terms of Eq. (A.22) by S_i and invoking the reciprocity principle, (A.14)-(A.15), we obtain a system of equations for the total fluxes incident on the surface

$$J_i - \sum_{k=1}^{r} \rho_k J_k \varphi_{ik} = \sum_{k=1}^{r} \mathcal{E}_{ck} \cdot S_{ik}, \quad (i = 1, 2, \ldots r) \tag{A.23}$$

having made use of the fact that

$$J_i = I_i S_i, \quad \mathcal{E}_{ck} = E_{ck} \cdot S_k, \text{ etc.} \tag{A.24}$$

b. The system of equations for the effective flux of molecules is obtained by substituting the value of I_i from (A.5) into (A.4):

$$E_e - \rho_i \sum_{k=1}^{r} E_{ek} \varphi_{ik} = E_{ci}. \tag{A.25}$$

From Eqs. (A.24) and the reciprocity principle we have

$$\mathcal{E}_{ei} - \rho_i \sum_{k=1}^{r} \mathcal{E}_{ek} \varphi_{ki} = \alpha_i \mathcal{E}_{0i}. \tag{A.26}$$

c. The system of equations for resultant absorption is derived by substituting the value of I_i according to (A.7) into (A.22):

$$A_{pi} - \alpha_i \sum_{k=1}^{r} \frac{\rho_k}{\alpha_k} \varphi_{ik} A_{pk} = \alpha_i \left(\sum_{k=1}^{r} E_{0k} \varphi_{ik} - E_{0i} \right), \tag{A.27}$$

and, accordingly, for the total resultant condensation

$$A_{pi} - \alpha_i \sum_{k=1}^{r} \frac{\rho_k}{\alpha_k} \varphi_{ki} A_{pk} = \alpha_i \sum_{k=1}^{r} \mathcal{E}_{ki} S_{ik}, \tag{A.28}$$

where

$$\mathcal{E}_{ki} = \mathcal{E}_{0k} = \mathcal{E}_{0i} \quad (i = 1, 2, \ldots r).$$

The finite systems of linear inhomogeneous algebraic equations obtained above, which may be used as the basis of molecular flow and mass transfer calculations in systems of bodies, are approximations of the corresponding integral equations (cf. Surinov, 1948).

The physical implication of this approximation is that, instead of an active emitting system of arbitrary configuration and dimensions with continuous fields of the concentrations and other physicochemical characteristics over its surface, we are dealing with a system comprising a finite number r of isothermal and homogeneous zones (or bodies) of ultimately simple configuration, meeting specific geometrical conditions. As the number r is increased the given system more nearly approaches the true system, coinciding with it in the limit as $r \to \infty$.

This is evident from the fact that from the physical point of view the integral equations of molecular flows may be interpreted as the equations of state of nonequilibrium emitting systems in which an interconnection is realized between different fields or factors of molecular transport.

To solve the problem as a whole it is sufficient to apply the corresponding integral molecular flux equation and thereby find the resultant field for any one type of flow. The resultant fields for other types of molecular flow are then determined by elementary algebraic transformations of the first solution, without having to bring in, and solve, the integral equations for the new problem.

If we permit simplifications in specifying the form of one or more of the fields (such as the temperature field, for example), we obtain commensurate simplifications in the functional equations of state. For instance, in the case of an emitting system consisting of r isothermal and homogeneous bodies of arbitrary configuration, the integral equations reduce to simpler finite systems of integral equations whose number corresponds to the number of zones delineated in the system.

If all components of the field are specified in simplified (rendered algebraic) form, and this is indeed possible if the configurations of the emitting system is such that the local angular coefficient $\phi(M_i S_k)$ for each zone S_i ($i = 1, 2, \ldots, r$) retains constant values at different points within the limits of this zone:

$$\varphi(M_i, S_k) = \text{const} \quad (i, k = 1, 2 \cdots r), \tag{A.29}$$

where ·

$$\varphi(M_i, S_k) = \frac{\partial \varphi_{ik}}{\partial S_i} / M_i,$$

then the integral equations degenerate into systems of algebraic equations which, under these conditions, are quite exact and rigorous.

We will present almost without derivation [this being apparent both from the text for the systems of algebraic equations and, for example, from the derivation of the integral equation (4.8)] the following:

Integral Equations for Various Elements of Molecular Flows

We give the following equations, assuming the law of cosines is valid.

a. The incident molecular flux on an area element dS_i about a point M_i on the surface of the i-th body is expressed by the equation

$$I(M) = \int_{(S_N)} \frac{E_e(N)}{\overline{\Omega}} \cos\theta_M d\Omega(MN) = \int_{(S_N)} E_e d\varphi(M,N), \tag{A.30}$$

where

$$d\varphi(M,N) = \frac{1}{\pi}\cos\theta(M)\, d\Omega(U,N) \tag{A.31}$$

is the elementary coupling coefficient for the areas dS_M and dS_N about the points M and N.

If there exists a system of r isothermal homogeneous bodies, the integral in (A.30) is conveniently written as a sum of integrals over the corresponding areas S_k:

$$I(M_i) = \sum_{k=1}^{r} \int_{(S_k)} E_e(N_k)\, d\varphi\,(M_i,N_k), \qquad (i = 1,2,\ldots r). \tag{A.32}$$

Substituting $E_e(N_k)$ into Eq. (A.32) according to the following:

$$E_e(N_k) = \alpha_k E_{ok} - \rho_k I(N_k), \tag{A.4a}$$

we obtain a system of integral equations for the incident molecular flux:

$$I(M_i) - \sum_{k=1}^{r} \rho_k \int_{(S_k)} I(N_k)\, d\varphi\,(M_i,N_k) = \sum_{k=1}^{r} \alpha_k E_{ok}\, \varphi(M_i,S_k), \tag{A.33}$$

and since

$$\varphi(M_i,S_k) = \int_{(S_k)} d\varphi(M_i,N_k) = \frac{1}{\pi}\int_{(S_k)} \frac{\cos\theta(M_i)\cos\theta(N_k)}{\rho^2_{M_iN_k}}\, dS_k \tag{A.34}$$

we obtain the system of integral equations for:

b. The effective molecular flux

$$E_e(M_i) - \rho_i \sum_{k=1}^{r} \int_{(S_k)} E_e(N_k)\, d\varphi(M_i,N_k) = \alpha_i E_{oi} \qquad (i = 1,2,\ldots r) \tag{A.35}$$

as well as:

c. The absorbed molecular flux, for example, by substitution of I_i according to Eq. (A.2) into the systems of equations (A.33):

$$\frac{A_n(M_i)}{\alpha_i} - \sum_{k=1}^{r} \frac{\rho_k}{\alpha_k} \int_{(S_k)} A_n(N_k)\, d\varphi(M_i,N_k) = \sum_{k=1}^{r} E_{ck}\, \varphi(M_i S_k). \tag{A.36}$$

It goes without saying that the systems of integral equations for the resultant condensation are obtained if we substitute A_a from (A.6b) into the system (A.36), while the system of integral equations for reflected fluxes is obtained by substituting I_i from Eq. (A.3) into (A.32).

The derived systems of integral equations occupy an intermediate status between the corresponding integral equations (for the continuous fields of such physicochemical properties as the temperature, micro- and macrogeometry, composition, crystallographic orientation, etc.) and the final systems of algebraic equations.

For example, the system of equations (A.33) is intermediate between the system of algebraic equations (A.22) and the integral equation (A.30).

The systems of integral equations go over to their algebraic counterparts if we assume that, for example, I, E_e, and the densities of other types of fluxes have a constant value within the confines of every region S_k. They can be obtained from the integral equations if we assume that the temperature and other characteristics are piecewise continuous functions on the surface of the system.

Let us apply the method of Surinov (1950) to establish the necessary and sufficient condition found by Polyak [1935 (1)] for a system of integral equations, say (A.35), to degenerate into a system of algebraic equations.

Multiplying the terms of the system (A.35) by dS_i and integrating over the surface S_i of the i-th zone, we obtain

$$\int_{(S_i)} E_e(M_i)\, dS_i - \rho_i \sum_{k=1}^{r} \int_{(S_k)} E_e(N_k)\, \varphi(N_k,S_k)\, dS_k = \alpha_i E_{0i} S_i \tag{A.37}$$

On the basis of (A.29) we rewrite the system of equation (A.37) in the form

$$\int_{(S_i)} E_e(M_i)\, dS_i - \rho_i \sum_{k=1}^{r} \varphi_{ki} \int_{(S_k)} E_e(N_k)\, dS_k = \alpha_i \mathcal{E}_{0i} \tag{A.37a}$$

The resultant equation may be rewritten in the form (A.26) and, on the basis of Eq. (A.24) and the reciprocity principle, we obtain

$$E_{ei} - \rho_i \sum_{k=1}^{r} \varphi_{ik} E_{ek} = \alpha_i E_{0i} \tag{A.15a}$$

Consequently, the condition (A.29) is the one we are seeking. The solution of the indicated systems of integral equations presents definite mathematical difficulties. They are amenable to solution essentially only with the application of modern computer techniques.

As an example, we give the problem of the molecular flow distribution in a cylinder with a bottom, treating the bottom of the cylinder as one zone, the walls as a second zone. The solution has been carried out in the work of Sparrow, Albers, and Eckert (1962) on radiative heat transfer.

Effusion from a Hollow Cylinder

Some of our earlier results (Secs. 4, 5, 7, 8) referred to a cylinder of sufficient length that the distribution of molecular flows in its cross section, i.e., perpendicular to the axis, could be neglected. It may also be assumed

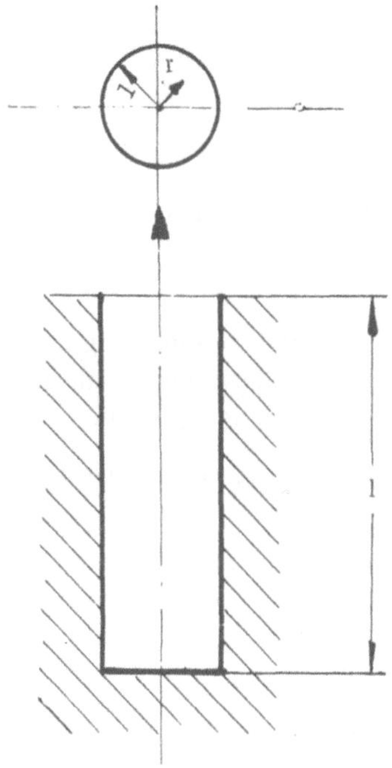

Fig. A.2. Geometry of the cylindrical cavity (Sparrow, Albers, and Eckert).

that the solutions obtained apply to the case when the dimensions of the molecular beam detector are of the same order of magnitude as the cross section of the investigated vessel. Under such circumstances, there would be no point in considering the variation in density of the molecules, for example, as a function of the coordinates in the plane of the exit opening.

The advantage of the solution presented below is that it takes into account irregularities of the molecular flow emanating from the bottom of a cylindrical cavity (Fig. A.2). This is essentially the same cylinder as in Fig. 9, but it is not taken for granted the desorption is uniform over the entire bottom.

It is intuitively understood that at the boundary of the walls and bottom of the cylinder the molecular flux density will be greater than along the axis of the cylinder. In order to obtain this as a quantitative result, we give the system of integral equations describing the effusion of monomers from the cylinder. We will relate the notation to the terminology of Section 7 and take into account irregularity of the emission from the bottom. We divide all terms of Eq. (7.1) by a, and for $\Phi(z_1)/a$ use the symbol $\bar{\Phi}(z_1)$, the dimensionless effective molecular flux, expressed in units of natural (Langmuir) emission from a point z_1 of the cylinder wall, $\bar{\mu}_M^D(r_1)$, the dimensionless molecular flux from a point r_1 on the bottom of the cylinder (the analog in terms of radiative heat transfer is the corresponding apparent emissivity).

If we assume that the condensation coefficient α is the same everywhere, we obtain the system of equations

$$\bar{\Phi}(z_1) = \alpha + (1-\alpha)\int_0^l \bar{\Phi}(z)\, d\varphi(z_1,z) + (1-\alpha)\int_0^l \bar{\mu}_M^D(r)\, d\varphi(z,r), \tag{A.38}$$

$$\bar{\mu}(r_1) = \alpha + (1-\alpha)\int_0^l \bar{\Phi}(z)\, d\varphi(r_1,z). \tag{A.39}$$

We need to establish the connection between $d\varphi(r_1, z)$, $d\varphi(z_1, z)$, $d\varphi(z_1, r)$ and the kernel of the integral equation (4.8) [whence was obtained, for example, Eq. (7.1)]. We will first show how $d\varphi(r_1, z)$ and $F(z_1, z)$ are related, then we will give the general formula. We make use of Fig. A.3. and apply elementary arguments of the type used in the derivation of Eqs. (4.7)-(4.8). To calculate the number of particles sent by a given ring $r-(r+dr)$ to a band $z-(z+dz)$ we first reckon how many particles are sent to this band from a disk of radius r, then from a disk of radius $r+dr$. The result will be the difference between these quantities:

$$d\varphi(r_1,z) = \frac{1}{4}\frac{\partial^2 F}{\partial r \partial z}\,. \tag{A.40}$$

For specific calculations we write Eq. (4.6) in the form

$$F(z,r) = z^2 + r^2 + 1 - \sqrt{(z^2+r^2+1)^2 - 4r^2}\,. \tag{A.41}$$

As an exercise the reader might carry out the operations indicated in (A.40) and obtain the specific values of $d\varphi(z_1, r)$, $d\varphi(r_1, z)$, and $d\varphi(z_1, z)$.

The results obtained by numerical solution of Eqs. (A.38)-(A.39) on a computer (by the method of successive approximations) are shown in Figs. A.4 and in Table A.2.

TABLE A.2. Effective Dimensionless Molecular Flux $\bar{\mu}_M^D(0)$ from the Center and $\bar{\mu}_M^D(1)$ from the Center of the Bottom of a Hollow Cylinder (SAE-Sparrow, Albers, and Eckert, 1962; B-Buckley, 1934)

l	$\alpha = 0.9$			$\alpha = 0.75$			$\alpha = 0.5$		
	$\bar{\mu}_M^D(0)$	$\bar{\mu}_M^D(1)$	$\bar{\mu}_M^D$	$\bar{\mu}_M^D(0)$	$\bar{\mu}_M^D(1)$	$\bar{\mu}_M^D$	$\bar{\mu}_M^D(0)$	$\bar{\mu}_M^D(1)$	$\bar{\mu}_M^D$
	SAE		B	SAE		B	SAE		B
8	0.9984	0.9986	0.99995	0.9956	0.9961	0.99975	0.9880	0.9887	0.9982
6	–	–	–	–	–	–	0.9768	0.9793	0.9924
4	0.9936	0.9945	0.9975	0.9815	0.9842	0.9921	0.9460	0.9540	0.9686
2	0.9785	0.9848	0.9850	0.9389	0.9553	0.9552	0.8394	0.8776	0.8714
1	0.9482	0.9708	0.9612	0.8626	0.9180	0.8938	0.6878	0.7914	0.7421
0.5	0.9191	0.9602	0.9377	0.7937	0.8904	0.8367	0.5693	0.7317	0.6383

In the columns under "B" in Table A.2 are given the values calculated by Buckley (1934) according to Eq. (7.12). It is readily apparent that the mean values of $\bar{\mu}_M^D$ according to Buckley for $l < 1$ lie between the values obtained by Sparrow, Alberts, and Eckert (SAE) for the edge and center of the disk. For $l \geq 1$ Buckley's data are too high in connection with the equation used. We note that the results of SAE can be used to determine the condensation coefficients α if the distribution of molecular beam density over the cross section of the exit aperture is measured. To do so, one must first obtain an expression for the total molecular efflux from the cylinder.

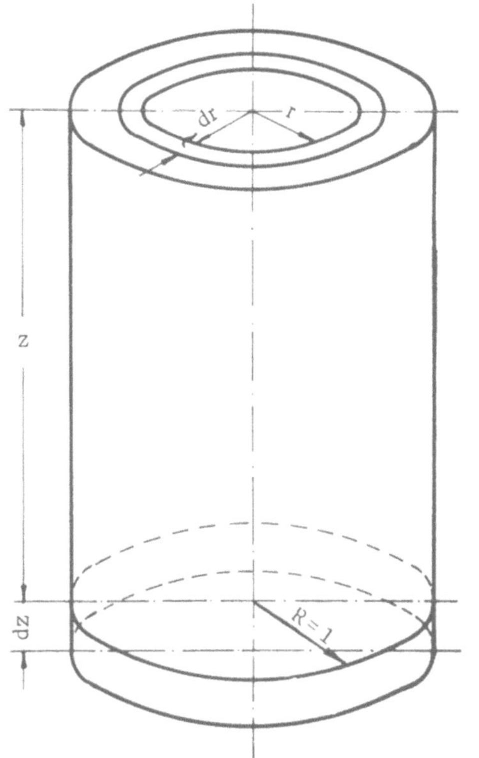

Fig. A.3. Diagram for calculation of the coupling coefficient between an elementary ring r-(r+dr) on the bottom of a cylinder and a band z-(z+dz) on the wall of the cylinder (Sparrow, Alberts, and Eckert).

We now give a certain generalization of Eq. (A.40).

It is understood that the kernel of the integral equation [in which collisions of particles in the volume, for example, of the type (A.38) are disregarded] should depend only on the form of the emission and scattering characteristics and the geometrical characteristics of the vessels, hence the following general expression is valid for the kernel:

$$F(\bar{x}_i, \bar{x}_k) \equiv \frac{f(\theta, \gamma, \bar{x}_k)\cos\theta_i}{\bar{\Omega}\,|\bar{x}_i - \bar{x}_k|^2} \tag{A.42}$$

where $\bar{\Omega}$ is the equivalent solid angle.

It can also be shown that the equation

$$F(\bar{x}_i, \bar{x}_k) = \frac{\partial^2 S_{ik}}{\partial S_i\, \partial S_k} \tag{A.43}$$

is valid, where S_{ik} is the mutual surface of the sections dS_i and dS_k situated at points with coordinates $\bar{x}_i$ and $\bar{x}_k$.

In the kernels of the integral equations in which processes of volume scattering and volume reactions are described, the kernel is complicated by a factor taking into account the probability of uneventful transfer from the origin of the particle to the point of observation in question.

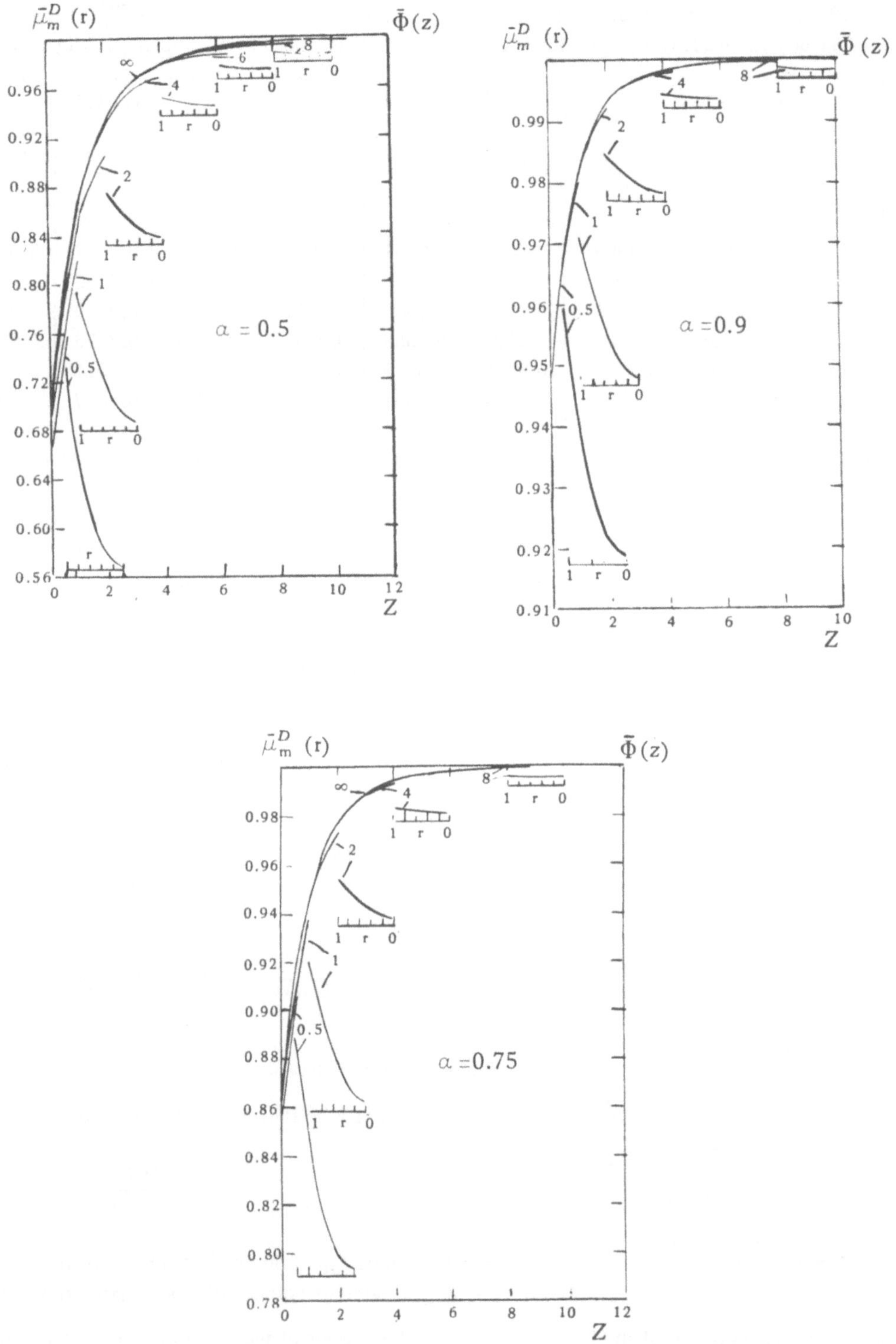

Fig. A.4. Lengthwise and radial distribution of effective molecular fluxes in a hollow cylinder for various condensation coefficients α. The numbers with the curves give the values of *l* (Sparrow, Alberts, and Eckert).

Description of the Phenomena in a More General Formulation of the Problem

We presupposed in the foregoing treatment that not only were the particles entering the molecular flows not subject to chemical conversion, but that they also had the same velocity, or we calculated certain integral characteristics of the flows wherein the velocities of individual particles were irrelevant.

In actuality there are certain functions describing the distribution of particles with respect to their energies and states. It can be shown rigorously that the number of possible states of particles, no matter how complicated the gas, form a denumerable set. This makes it possible to treat the gas as a mixture of particles of different types i, corresponding to the set of possible states i.

Knowing the set of distribution functions $f_i(\bar{x}, \bar{u}, t)$, which are numbered in accordance with the numbering of the states or types of gas particles, we can describe the state of the gas (f_i is not to be confused with the directional characteristic).

Let the position of the ith type of particle (we will occasionally refer simply to the "i-the particle") be specified by a set of three numbers x_1, x_2, x_3, or, in short form, $\bar{x}$, with corresponding particle velocities $\bar{u}$; then the probability dn_i of one particle being incumbent at the time t in the volume $d\bar{x}d\bar{u}$ of phase space between the points

$$\begin{aligned} &x_1 - x_1 + dx_1\,, \\ &x_2 - x_2 + dx_2\,, \\ &x_3 - x_3 + dx_3\,, \\ &u_1 - u_1 + du_1\,, \\ &u_2 - u_2 + du_2\,, \\ &u_3 - u_3 + du_3\,. \end{aligned} \tag{A.44}$$

is given by the expression

$$dn_i = f_i\,(\bar{x},\bar{u},t)\,d\bar{x}\,d\bar{u}. \tag{A.45}$$

The function f_i has a twofold sense: the mathematical expectation density of the number of particles in six-dimensional space and the probability density of a single i-th particle being found in six-dimensional space. Similarly, dn_i has the sense of the mathematical expectation of the number of particles in the indicated element of phase space.

The distribution function introduced above is sufficient for describing any system of particles, but for convenience of presentation we will divide f_i into several types, corresponding to the types of particle phase state.

Let us distinguish i-th particles situated in a gaseous volume, those adsorbed on a crystallographically smooth surface, those on crystal steps or grain boundaries, and those inside the crystal. Consequently,

$$f_i = f_{ik} \quad (k = 0,1,2,3). \tag{A.46}$$

For definiteness, k = 0 will refer to states in the crystal lattice, k = 1 to adsorbtion on a step, k = 2 to adsorption on a surface, k = 3 to a gaseous volume. The number of possible k could be more than four, and their physical interpretation could prove to be other than as we have described it, but we will not be concerned about this, at least for the moment.* Another factor that simplified the problem was the assumption that collisions did not occur in the gaseous volume. We will cast off this limitation as well and consider the case when there is an appreciable probability of collision between two particles in the gaseous volume. The mean free path will then be decreased to fractions of the transverse dimensions of the cavity in which the gas is contained.

Among the reactions to which the particles are subjected we cite monomolecular reactions, spontaneous and induced dissociation, volume reactions, bimolecular reactions (with increasing pressure), etc.

*For example, sometimes the subscript "O" conveniently denotes vacancies on the surface.

Table A3. Probability That a Particle of Type pi Will Not Once Collide with Particles of Type mk in the Time (τ, t); Table Based on Eq. (A52'):

$$N_{pimk}(\bar{x}_i, \bar{u}_i, \tau, t) = \exp\left\{\sum_{mk} - \int_{\tau}^{t}\int_{-\infty}^{\infty} |\bar{u}_{pi}(\tau') - \bar{u}_{mk}|\,\sigma_{pimk}\, f_{mk}\, d\bar{u}_{mk}\, d\tau'\right\} \quad (A.52')$$

Probability of no collision of gas particles i and k.	p = 3 m = 3	$N_{3i3k} = \exp\left\{\sum_{3k} - \int_{\tau}^{t}\int_{-\infty}^{\infty} \lvert\bar{u}_{3i}(\tau') - \bar{u}_{3k}\rvert\,\sigma_{3i3k}\, f_{3k}\, d\bar{u}_{3k}\, d\tau'\right\}$
Probability of no collision of gas particles i with particles k adsorbed on the surface.	p = 3 m = 2	$N_{3i2k} = \exp\left\{\sum_{2k} - \int_{\tau}^{t}\int_{-\infty}^{\infty} \lvert\bar{u}_{3i}(\tau') - \bar{u}_{2k}\rvert\,\sigma_{3i2k}\, f_{2k}\, d\bar{u}_{2k}\, d\tau'\right\}$
Probability of no collision of gas particles i with particles k adsorbed on a step.	p = 3 m = 1	$N_{3i1k} = \exp\left\{\sum_{1k} - \int_{\tau}^{t}\int_{-\infty}^{\infty} \lvert\bar{u}_{3i}(\tau') - \bar{u}_{1k}\rvert\,\sigma_{3i1k}\, f_{1k}\, d\bar{u}_{1k}\, d\tau'\right\}$
Probability of no collision of gas particles i with particles k in the crystal.	p = 3 m = 0	$N_{3i0k} = \exp\left\{\sum_{0k} - \int_{\tau}^{t}\int_{-\infty}^{\infty} \lvert\bar{u}_{3i}(\tau') - \bar{u}_{0k}\rvert\,\sigma_{3i0k}\, f_{0k}\, d\bar{u}_{0k}\, d\tau'\right\}$
Probability of no collision of particles i and k adsorbed on the surface.	p = 2 m = 2	$N_{2i2k} = \exp\left\{\sum_{2k} - \int_{\tau}^{t}\int_{-\infty}^{\infty} \lvert\bar{u}_{2i}(\tau') - \bar{u}_{2k}\rvert\,\sigma_{2i2k}\, f_{2k}\, d\bar{u}_{2k}\, d\tau'\right\}$

Table A.3 (cont.)

Probability of no collision of particles i adsorbed on the surface with particles k adsorbed on a step.	p = 2 m = 1	$N_{2i1k} = \exp\left\{\sum_{1k} - \int_{\tau}^{t}\int_{-\infty}^{\infty} \lvert \bar{u}_{2i}(\tau') - \bar{u}_{1k} \rvert \sigma_{2i1k} f_{1k}\, d\bar{u}_{1k}\, d\tau'\right\}$
Probability of no collision of particles i adsorbed on the surface with particles k in the crystal.	p = 2 m = 0	$N_{2i0k} = \exp\left\{\sum_{0k} - \int_{\tau}^{t}\int_{-\infty}^{\infty} \lvert \bar{u}_{2i}(\tau') - \bar{u}_{0k} \rvert \sigma_{2i0k} f_{0k}\, du_{0k}\, d\tau'\right\}$
Probability of no collision of particles i and k adsorbed on a step.	p = 1 m = 1	$N_{1i1k} = \exp\left\{\sum_{1k} - \int_{\tau}^{t}\int_{-\infty}^{\infty} \lvert \bar{u}_{1i}(\tau') - \bar{u}_{1k} \rvert \sigma_{1i1k} f_{1k}\, d\bar{u}_{1k}\, d\tau'\right\}$
Probability of no collision of particles i adsorbed on a step with particles k in the crystal.	p = 1 m = 0	$N_{1i0k} = \exp\left\{\sum_{0k} - \int_{\tau}^{t}\int_{-\infty}^{\infty} \lvert \bar{u}_{1i}(\tau') - \bar{u}_{0k} \rvert \sigma_{1i0k} f_{0k}\, d\bar{u}_{0k}\, d\tau'\right\}$
Probability of no collision of particles i and k inside the crystal.	p = 0 m = 0	$N_{0i0k} = \exp\left\{\sum_{0k} - \int_{\tau}^{t}\int_{-\infty}^{\infty} \lvert \bar{u}_{0i}(\tau') - \bar{u}_{0k} \rvert \sigma_{0i0k} f_{0k}\, d\bar{u}_{0k}\, d\tau'\right\}$

Particles of the i-th type, impinging in the volume of the reactor, can either collide with another particle or with the surface of the reactor. In the case of collision with a particle it can change its velocity and state (becoming dissociated or bound with this particle, or a fragment of it becoming bound with the particle, etc.). If it impinges on the surface, there are again many possibilities, depending on the state of the site of incidence, and what strikes what.

The chain of events turns out to be rather complicated and protracted. In order to isolate the important facts, we will carry out an analysis using the results obtained by Vallander (1960-64) and co-workers. The publications of the Vallander school pertain to the aerodynamics of rarefied gases and are not immediately concerned with the problems of molecular beam transmission through the vessels confining them. However, some of their results may be applied to our own area of interest, and in fact we have done this in the ensuing text.

Probability of Uneventful Transfer

We use the expression "uneventful" not only to imply absence of spontaneous or induced conversions of the kth type, but also the absence of collision between particles of the same or different types with changes in their velocities (either in magnitude or direction). We will assume for greater generality (from which we can narrow down to the simpler special case if so desired) that the gas is situated in a constant external force field with acceleration constant $\bar{g}$. Then the radius vector $\bar{x}_i$ defining the position of the i-th particle at any time τ' ($\tau \leq \tau' \leq t$) and the velocity $\bar{u}_i$ of this particle may be written in the form

$$\bar{x}_i(\tau') = \bar{x}_i - \bar{u}_i(t - \tau') + \frac{\bar{g}(t-\tau')^2}{2}, \tag{A.47}$$

$$\bar{u}_i(\tau') = \bar{u}_i - \bar{g}(t - \tau'). \tag{A.48}$$

The collision cross section of particles of type i and k is said to be σ_{ik}; it depends on the relative velocity of the colliding particles.

We will say for simplicity that $\sigma_{ik} = \sigma_{ik}(|\bar{u}_i - \bar{u}_k|)$. It is well known from quantum mechanics that $\sigma_{ik} = \sigma_{ki}$. We use dQ_{ik} to denote the probability of a random collision during the time $d\tau$ of an i-th particle with a k-th particle. This collision will happen if in the volume

$$dV = \sigma_{ik}(|\bar{u}_i - \bar{u}_k|)\,|\bar{u}_i - \bar{u}_k|\,d\tau \tag{A.49}$$

described by σ_{ik} and moving with a relative velocity $|\bar{u}_i - \bar{u}_k|$ during the time $d\tau$ there turns out to be at least one atom with velocity $\bar{u}_k$ in the interval $\bar{u}_k$ to $(\bar{u}_k + d\bar{u}_k)$.

Then

$$dQ_{ik} = f_k(\bar{x}_k + \bar{u}_k\tau', \bar{u}_k, t + \tau')\,\sigma_{ik}\,|\bar{u}_i - \bar{u}_k|\,d\bar{u}_k d\tau\,. \tag{A.50}$$

The probability Q_{ik} that the i-th particle situated at the point $\bar{x}_i(\tau')$ and having the velocity $\bar{u}_i(\tau')$, at the time τ', will collide with some particle in the time $d\tau$ may be written in the form

$$Q_{ik} = d\tau \int_{-\infty}^{\infty} |\bar{u}_i(\tau) - \bar{u}_k|\,\sigma_{ik} f_k(\bar{x}_k, \bar{u}_k, \tau')\,d\bar{u}_k\,, \tag{A.51}$$

and the probability N_{ik} that the i-th particle will not once collide during the period τ to t with particle of one definite type k is expressed in the form

$$N_{ik}(\bar{x}_i, \bar{u}_i, \tau, t) = \exp\left\{-\iint_{\tau\,-\infty}^{t\,\infty} |\bar{u}_i(\tau) - \bar{u}_k|\,\sigma_{ik}(|\bar{u}_i(\tau') - \bar{u}_k|)\,f_k[\bar{x}_i(\tau'), \bar{u}_k, \tau']\,d\bar{u}_k dt\right\}\,. \tag{A.52}$$

Table A4. Probability $N_{pi}(\bar{x}_{pi}, \bar{u}_{pi}, \tau, t)$ of Uneventful Motion of Particles of Type pi During the Period τ-t; Table Based on the Equation:

$$N_{pi}(\bar{x}_{pi}, \bar{u}_{pi}, \tau, t) = exp\left\{-\int_{\tau}^{t}\left[\sum_{mk}\int_{-\infty}^{\infty}|\bar{u}_{pi}(\tau') - \bar{u}_{mk}|\sigma_{pimk} f_{mk}\left(\bar{x}_{pi}(\tau'), \bar{u}_{mk}, \tau'\right) d\bar{u}_{mk} + \int_{-\infty}^{\infty} P_{pi}(\bar{x}_{pi}, \bar{u}_{pi}, \tau, t) f_{pi}\, d\bar{u}_{pi}\right] d\tau'\right\}$$

Probability of particle moving unimpeded in a gaseous volume.	p = 3	$N_{3i} = exp\left\{-\int_{\tau}^{t}\left[\sum_{mk}\int_{-\infty}^{\infty}\lvert\bar{u}_{3i} - \bar{u}_{mk}\rvert \sigma_{3imk} f_{mk} d\bar{u}_{mk} + \int_{-\infty}^{\infty} P_{3i} f_{3i} d\bar{u}_{3i}\right] d\tau'\right\}$
Probability of particle moving unimpeded on a surface.	p = 2	$N_{2i} = exp\left\{-\int_{\tau}^{t}\left[\sum_{mk}\int_{-\infty}^{\infty}\lvert\bar{u}_{2i} - \bar{u}_{mk}\rvert \sigma_{2imk} f_{mk} d\bar{u}_{mk} + \int_{-\infty}^{\infty} P_{2i} f_{2i} d\bar{u}_{2i}\right] d\tau'\right\}$
Probability of particle moving unimpeded on a step.	p = 1	$N_{1i} = exp\left\{-\int_{\tau}^{t}\left[\sum_{mk}\int_{-\infty}^{\infty}\lvert\bar{u}_{1i} - \bar{u}_{mk}\rvert \sigma_{1imk} f_{mk} d\bar{u}_{mk} + \int_{-\infty}^{\infty} P_{1i} f_{1i} d\bar{u}_{1i}\right] d\tau'\right\}$
Probability of particle moving unimpeded in the crystal interior.	p = 0	$N_{0i} = exp\left\{-\int_{\tau}^{t}\left[\sum_{mk}\int_{-\infty}^{\infty}\lvert\bar{u}_{0i} - \bar{u}_{mk}\rvert \sigma_{0imk} f_{mk} d\bar{u}_{mk} + \int_{-\infty}^{\infty} P_{0i} f_{0i} d\bar{u}_{0i}\right] d\tau'\right\}$

Table A5. Probability of Collision of Particles of Type mk and nl in a Unit Phase Volume per Unit Time; Table Based on Eq. (A56'):

$$E_{mknl} = |\bar{u}_{mk} - \bar{u}_{nl}|\, \sigma_{mknl}(|\bar{u}_{mk} - \bar{u}_{nl}|)\, f_{mk}(\bar{x}_{mk}, \bar{u}_{mk}, t)\, f_{nl}(\bar{x}_{nl}, \bar{u}_{nl}, t)$$

Collision of particles k and l in a gaseous volume.	m = 3 n = 3	$E_{3k3l} = \lvert\bar{u}_{3k} - \bar{u}_{3l}\rvert\, \sigma_{3k3l}\, f_{3k} f_{3l}$
Collision of gas particles k with particles l adsorbed on the surface.	m = 3 n = 2	$E_{3k2l} = \lvert\bar{u}_{3k} - \bar{u}_{2l}\rvert\, \sigma_{3k2l}\, f_{3k} f_{2l}$
Collision of gas particles k with particles l adsorbed on a step.	m = 3 n = 1	$E_{3k1l} = \lvert\bar{u}_{3k} - \bar{u}_{1l}\rvert\, \sigma_{3k1l}\, f_{3k} f_{1l}$
Collision of gas particles k with particles l in the crystal.	m = 3 n = 0	$E_{3k0l} = \lvert\bar{u}_{3k} - \bar{u}_{0l}\rvert\, \sigma_{3k0l}\, f_{3k} f_{0l}$
Collision of particles k and l adsorbed on the surface.	m = 2 n = 2	$E_{2k2l} = \lvert\bar{u}_{2k} - \bar{u}_{2l}\rvert\, \sigma_{2k2l}\, f_{2k} f_{2l}$
Collision of particles k adsorbed on the surface with particles l adsorbed on a step.	m = 2 n = 1	$E_{2k1l} = \lvert\bar{u}_{2k} - \bar{u}_{1l}\rvert\, \sigma_{2k1l}\, f_{2k} f_{1l}$
Collision of particles k adsorbed on the surface with particles l in the crystal.	m = 2 n = 0	$E_{2k0l} = \lvert\bar{u}_{2k} - \bar{u}_{0l}\rvert\, \sigma_{2k0l}\, f_{2k} f_{0l}$
Collision of particles k and l adsorbed on a step.	m = 1 n = 1	$E_{1k1l} = \lvert\bar{u}_{1k} - \bar{u}_{1l}\rvert\, \sigma_{1k1l}\, f_{1k} f_{1l}$
Collision of particles k adsorbed on a step with particles l in the crystal.	m = 1 n = 0	$E_{1k0l} = \lvert\bar{u}_{1k} - \bar{u}_{0l}\rvert\, \sigma_{1k0l}\, f_{1k} f_{0l}$
Collision of particles k and l in the crystal interior.	m = 0 n = 0	$E_{0k0l} = \lvert\bar{u}_{0k} - \bar{u}_{0l}\rvert\, \sigma_{0k0l}\, f_{0k} f_{0l}$

In view of the fact that we have introduced functions of the type f_{pi} (p = 0, 1, 2, 3,), Eq. (A.52) can be used to generate Table (A.3).

The probability of free motion of an i-th particle during the period τ - T, when the position and velocity of the particle at the time τ' are determined by the expressions (A.47)-(A.48), is determined, according to the probability multiplication theorem, by the equation

$$N_i^0(\bar{x}_i,\bar{u}_i,\tau,t) = \exp\left\{\sum_k -\int_\tau^t\int_{-\infty}^{\infty} |\bar{u}_i(\tau') - \bar{u}_k| \, \sigma_{ik} f_k(\bar{x}_i(\tau'), \bar{u}_k, \tau')\, d\bar{u}_k d\tau'\right\}. \quad (A.53)$$

We introduce the probability $P_{pi}(\bar{x}_{pi}, \bar{u}_{pi}, \tau, t)$ of spontaneous disintegration during the time τ - t of a particle of type pi, where p = 0, 1, 2, 3, situated at a point with coordinates $\bar{x}_{pi}$ and having the velocity $\bar{u}_{pi}$. The velocity dependence is eliminated in a number of cases, but, for example, in the case of spontaneous thermal disintegration of complex particles, it is possible for P_{pi} to be a function of the temperature and, hence, of the velocity.

The probability N'_{pi} of an i-th particle moving during a time τ - t without spontaneous disintegration is determined by the expression

$$N'_{pi} = \exp\left\{-\int_\tau^t\int_{-\infty}^{\infty} P_{pi} f_{pi}\, d\bar{u}_{pi} d\tau'\right\}. \quad (A.54)$$

It is clear that the probability N'_{pi} of an uneventful transfer is determined by the product

$$N_{pi} = N^0_{pi} \cdot N'_{pi}, \quad (A.55)$$

from which we are able, as with Eq. (A.52) and Table A.3, to build Table A.4.

Particle Production

Particles of type pi may appear either spontaneously or upon collision between particles of type mk and n*l*. Here m and n, like p, are equal to 0, 1, 2, or 3 and indicate the position of particles of type i, k, *l*.

It is clear from Eq. (A.50) with $\tau' = 0$ that the number of collisions of particles of type k and *l* during the time dt in the element $d\bar{x}$ is given by the equation

$$E_{mknl}\, dt\, dx\, d\bar{u}_{mk}\, d\bar{u}_{nl} = |\bar{u}_{mk} - \bar{u}_{nl}|\, \sigma_{mknl} f_{mk} f_{nl}\, dt\, d\bar{x}\, d\bar{u}_{mk} d\bar{u}_{nl}\ . \quad (A.56)$$

For functions of the type E_{3k2l}, E_{3k1l}, E_{3k0l}, which refer to the probabilities of collisions of gaseous particles k with particles *l* on the surface (under different energy and geometrical conditions), Eq. (A.56) can be used quite formally, and a notation that is sometimes more suitable can be introduced.

Let the points $\bar{x}_S$ lie on the surface, whereupon the number of collisions of particles with an element of surface dS in the time dt is determined by the expression

$$E_{3knl} = f_{3k}(\bar{x}_s,\bar{u}_k,t)\, |\bar{u}_{kN}|\, dt\, dS\, d\bar{u}_k\ , \quad (A.57)$$

where $\bar{u}_{kN}$ is the component of $\bar{u}_k$ normal to the surface. We might be interested in the collision of particles adsorbed on portions of a crystallographically flawless surface or with a plane surface with steps intersecting the surface. In this and in other situations it is necessary to know the distribution functions of the various kinds of steps on the surface (cf. Chernov, 1957).

We note once again that particles of type Ok may include vacancies for particles of type i. For a complete description of the surface it is useful to know the conditions anent these vacancies (velocity, distribution on the surface at a given instant).

In this context we present Table A.5, which gives the functions E_{mknl}.

Once we have obtained the equations for calculating the number of particle collisions in an element of phase volume during the time dt, we should be able to calculate th birthrate of type i particles as a result of collision. But to do this we must know the outcome of the collision or, in other words, the probability $T^{pi}_{mknl}(\bar{u}_{mk}, \bar{u}_{nl}, \bar{u}_{pi})\, d\bar{u}_{pi}$ of obtaining particles of type pi when it is certain that particles of type mk and nl will collide.

As already stated, particles of type pi may also appear as the result of spontaneous disintegration of, for example, particles of type mk. The probability of a particle of type pi appearing when the disintegration of particles of type mk is certain is denoted by $P^{pi}_{mk}(\bar{u}_{mk}, \bar{u}_{pi})\, d\bar{u}_{mk}$.

We could also present tables for the functions T^{pi}_{mknl} and P^{pi}_{mk}, but this is something which the reader can do on his own.

We point out that, for example, P^{3i}_{mk}, where m = 0, 1, 2 is associated with Langmuir vaporization or desorption from different energy levels.

It is important to state that some of the probabilities derived herein are clearly equal to 0, for example, $P^{pi}_{3k} = 0$ for p = 0, 1, 2, since this production is not spontaneous. The reader will be able to deduce the exclusion principles from the specific conditions of the problem to be solved.

Some Specific Values of Functions of the Type T^{pi}_{mknl}

In this section we will present some results of calculations according to specific models of collisions of particles with other particles and with surfaces. All of the results published in the collection by Vallander (Aerodynamics of Rarefied Gases, Collection 1, 1963) will be indicated in the form (author, V1, page no.), those in Collection 2 of the same work (1965) in the form (author, V2, page no.).

In these works are calculated functions of the type T^{3i}_{3i3i}, T^{3i}_{3inp} (n = 0, 1, 2). The laws of conservation can be used as a basis for deriving several normalized relations (Vallander, V1, 7):

$$\int_{-\infty}^{\infty} T^{3i}_{3i3k}(\bar{u}_i, \bar{u}_k, \bar{u})\, d\bar{u} = 2 \,, \tag{A.58}$$

$$\int_{-\infty}^{\infty} \bar{u}\, T^{3i}_{3i3i}(\bar{u}_1, \bar{u}_2, \bar{u})\, d\bar{u} = \bar{u}_1 + \bar{u}_2 \,, \tag{A.59}$$

$$\int_{-\infty}^{\infty} \bar{u}^2 T^{3i}_{3i3i}(\bar{u}_1, \bar{u}_2, \bar{u})\, d\bar{u} = \bar{u}_1^2 + \bar{u}_2^2 \,, \tag{A.60}$$

$$\int_{(u_N > 0)} T^{3i}_{3ini}\, d\bar{u} = \rho \qquad n \neq 3 \,, \tag{A.61}$$

where ρ is the reflection coefficient, u_N is the projection of the velocity on the normal to the surface area.

Let us examine Eq. (A.58) in more detail (Vallander, V1, 80). Suppose that we have a gas with particles (masses m_1 and m_2) moving with velocities $\bar{u}_1$, $\bar{u}_2$ before impact and $\bar{u}_1'$, $\bar{u}_2'$ after impact. Then their relative velocity $\bar{u}_0$ is determined by the equation

$$\bar{u}_0 = \bar{u}_2 - \bar{u}_1 \tag{A.62}$$

and the velocity of the center of gravity $\bar{u}_c$ is determined by the formula

$$\bar{u}_c = \frac{m_1\bar{u}_1 + m_2\bar{u}_2}{m_1 + m_2}, \tag{A.63}$$

whence

$$\bar{u}_1 = \bar{u}_c - \frac{m_2}{m_1 + m_2}\bar{u}_0 ,$$
$$\bar{u}_2 = \bar{u}_c + \frac{m_1}{m_1 + m_2}\bar{u}_0 . \tag{A.64}$$

From the conservation of momentum ($\bar{u}_c = \text{const}$) we obtain

$$(\bar{u}_1 - \bar{u}_c)^2 = (\bar{u}_1' - \bar{u}_c)^2 = \frac{m_2^2}{(m_1 + m_2)^2}|\bar{u}_0|^2 = R_1^2 ,$$

$$(\bar{u}_2 - \bar{u}_c)^2 = (\bar{u}_2' - \bar{u}_c)^2 = \frac{m_1^2}{(m_1 + m_2)^2}|\bar{u}_0|^2 = R_2^2 . \tag{A.65}$$

We introduce the spherical coordinates R, ϑ, φ with center at the point u_{cx}, u_{cy}, u_{cz} and axis $\vartheta = 0$ in the direction of $\bar{u}_0 = \bar{u}_2 - \bar{u}_1$. Let

$$T_{3i3k}^{3k}(\bar{u}_i, \bar{u}_k, \bar{u})\, d\bar{u} = T_2(\bar{u}_1, \bar{u}_2, \bar{u})\, d\bar{u} \tag{A.66}$$

be the probability that after collision of two particles with velocities $\bar{u}_1$ and $\bar{u}_2$ the second will have a velocity somewhere in the element $d\bar{u}$, which contains the end of the vector $\bar{u}$.

$$T_2(\bar{u}_1, \bar{u}_2, \bar{u})\, d\bar{u} = T_2(\bar{u}_1, \bar{u}_2, R, \vartheta, \varphi)\, R^2 \sin\vartheta\, dR\, d\vartheta\, d\varphi$$

for $R \neq R_2$, by virtue of (A.65) $T = 0$.

This means that we may write the following, bringing in the Dirac delta function δ:

$$T_2 = K\, T_\Omega(\vartheta, \varphi)\, \delta(R - R_2) .$$

We normalize the function T_Ω with respect to the solid angle of scattering:

$$\int T_\Omega\, d\Omega = \int_0^{2\pi}\int_0^{\pi} T_\Omega(\vartheta, \varphi)\sin\vartheta\, d\vartheta\, d\varphi = 1.$$

K is determined from a condition of the type (A.58), i.e., the probability of $\bar{u}_2'$ existing somewhere in all of velocity space is equal to unity:

$$\int_{-\infty}^{\infty} T_2\, d\bar{u} = K\int_\Omega T_\Omega\, d\Omega \int_0^{\infty} R^2\delta(R - R_2)\, dR = KR_2^2 = 1 .$$

Hence

$$T_2 = \frac{1}{R_2^2} T_\Omega(\vartheta, \varphi)\,\delta(R - R_2).$$

In analogous fashion we also derive an expression

$$T_1 = \frac{1}{R_1^2} T_\Omega(\pi - \vartheta, \pi + \varphi)\,\delta(R - R_1)\ .$$

We now introduce the reduced mass

$$M = \frac{m_1 m_2}{m_1 + m_2}$$

and, noting that ϑ is the angle between $\bar{u}_1 - \bar{u}_c$ and $\bar{u}_2 - \bar{u}_c$, we obtain

$$T = \frac{1}{2}(T_1 + T_2) = \frac{1}{2M^2 |\bar{u}_0|^2}\,[m_1^2 T_\Omega(\pi - \vartheta, \pi + \varphi)\,\delta(R - R_1) + + m_2^2 T_\Omega(\vartheta, \varphi)\,\delta(R - R_2)] \tag{A.67}$$

For the case of inelastic interaction

$$T = \frac{1}{2}\sum_{i=1}^{2}\sum_{n}\frac{p_i^{(n)}}{R_i^{(n)2}} T_{\Omega i}^{(n)}(\vartheta, \varphi)\,\delta(R - R_i^{(n)}), \tag{A.68}$$

where $p_i^{(n)}$ is the probability of transition of an i-th particle (i = 1, 2) to the n-th energy level.

The scattering function T_Ω bears information as to the nature and mechanism of the collision.

For a central collision, according to classical mechanics (Mott and Massey, 1951)

$$T_\Omega = -\frac{s\,ds}{\sigma \sin\vartheta\, d\vartheta} \tag{A.69}$$

where s is the limiting distance, ϑ is the angle of deflection, σ is the collision cross section.

If s is not unique, then Eq. (A.69) is written in the form

$$T_\Omega(\vartheta) = \frac{1}{\sigma \sin\vartheta}\sum_i s_i \left|\frac{ds_i}{d\vartheta}\right| \tag{A.70}$$

T_Ω is determined by the interaction potential U(r), where r is the distance between particles.

Several models have been analyzed to date (see Hirschfelder, Curtiss, and Bird, 1961). We cite as an example the case of elastic spheres:

$$U(r) = \begin{cases} 0 & r > r_0 \\ \infty & r \le r_0 \end{cases} .$$

For $\vartheta = 0$; $\sigma = \pi r_0^2$,

$$s(\vartheta) = r_0 \cos\frac{\vartheta}{2};\ T_\Omega = \frac{1}{4\pi} = \text{const.},$$

$$T(\bar{u}_1,\bar{u}_2,\bar{u}) = \frac{1}{\pi|\bar{u}_2 - \bar{u}_1|^2}\,\delta\left\{\left|\bar{u} - \frac{\bar{u}_1 + \bar{u}_2}{2}\right| - \frac{|\bar{u}_2 - \bar{u}_1|}{2}\right\}.$$

The following relations have more general application:

$$\pi - \vartheta = 2\int_{r_m}^{\infty} \frac{s\,|\bar{u}_0|\,dr}{r^2\sqrt{|\bar{u}_0|^2 - \frac{2}{M}U(r) - \frac{s^2|\bar{u}_0|^2}{r^2}}},$$

$$|\bar{u}_0|^2 - \frac{2}{M}U(r_m) - \frac{s^2|\bar{u}_0|^2}{r_m^2} = 0. \qquad \text{(A.71)}$$

The integral (A.71) has been tabulated in the work of Hirschfelder, Curtiss, and Bird (1961).

The calculation of functions of the type T^{3i}_{3knl} is rather complex and has not yet been solved in general form. For equilibrium it corresponds to diffuse scattering at a surface and, consequently, does not depend on either the energy or the direction of the particles 3k, only on the temperature of the surface (particles nl).

Then the probability of a particle escaping with a velocity u at an angle ϑ with respect to the normal $\bar{N}$ in the solid angle $d\Omega = \sin\vartheta\, d\vartheta\, d\gamma$ is determined by the formula

$$\int_{u=0}^{\infty} T^{3i}_{ni}\, u^2\, du\, d\Omega = \frac{\cos\vartheta}{\pi}\, d\Omega \quad n = 0,1,2 \qquad \text{(A.72)}$$

$$T^{3i}_{ni}(\bar{N},\vartheta,\bar{u}) = \frac{m^2}{2(kT)^2}\, u_N e^{-mu^2/2kT}. \qquad \text{(A.73)}$$

When equilibrium is absent the law of cosines is not valid in general.

The law governing scattering of particles arriving at the surface depends on the structure of the surface and a host of other factors. Barantsev (A1, 107) has found the law of particle scattering from a "normal" surface, whose deviations from the plane obey a normal distribution law.

A particle impinging on such a "normal" surface at a velocity $\bar{u}_1$ has a probability $T_1(\bar{u}_1, \bar{u})$ of escaping from this surface at a velocity $\bar{u}$:

$$\tilde{T}_1(\bar{u}_1,\bar{u}) = \frac{1}{2\pi\sigma_1^2} \int\limits_{\alpha_1 < \pi/2} T_0(\bar{u}_1, N, \bar{u})\, e^{-\tan^2\vartheta/2\sigma_1^2} \frac{\cos\alpha_1\, d\Omega}{\cos\theta_1 \cos 4\vartheta} =$$
$$= \frac{1}{2\pi\sigma_1^2} \int\limits_{\alpha_1 < \pi/2} T_0 e^{-\tan^2\vartheta/2\sigma_1^2} (1 + \tan\theta_1 \tan\vartheta \sin\gamma) \frac{d\Omega}{\cos^3\vartheta}, \tag{A.74}$$

where ϑ is the angle between the normal to the scattering microsurface and the normal to the macroplane, θ_1 is the angle between the direction of the particle and the normal to the macroplane, $\sigma_1 = \sqrt{2}\,\sigma/\rho$, where σ is the surface fluctuation, ρ is the radius of correlation for the surface, γ is the macroazimuth, α_1 is the angle of incidence on the scattering microsurface, T_0 is the scattering law on the microsurface, $\bar{N}$ is the normal to the microsurface.

If $T_0(\bar{u}_1, \bar{N}, \bar{u})$ is determined, for example, by Eq. (A.73):

$$T_0 = \frac{m^2}{2k^2T^2}\, u \cos\alpha\, e^{-mu^2/2kT}$$

then after appropriate calculation we obtain

$$T_1 = \frac{m^2}{2k^2T^2\pi}\, u\cos\theta\, e^{-mu^2/2kT} \{1 - g_1 + g_2 \tan\theta_1 \tan\vartheta \sin\varphi, \tag{A.75}$$

where θ is the angle between the direction of escape and the macronormal,

$$g_1 = 1 - \sqrt{\frac{\pi}{2}}\frac{1}{\sigma_1} e^{1/2\sigma_1^2}\left[1 - \varphi\left(\frac{1}{\sigma_1\sqrt{2}}\right)\right],$$

$$g_2 = \frac{1}{2}\{g_1 + \sigma_1^2 [1 - g_1]\}, \tag{A.76}$$

and

$$\varphi(x) = \frac{2}{\sqrt{\pi}} \int_0^x e^{-t^2} dt .$$

If the total reflection due to the microsurfaces forming the macrosurface obeys the law T_0, then for small σ_1 (small roughness scale)

$$T_1 = T_0 [1 + g \tan\vartheta_1 \tan\vartheta \sin\varphi] . \tag{A.77}$$

where

$$g = \frac{g_2}{1 - g_1} \tag{A.78}$$

and

σ_1	0.1	0.2	0.33
g	0.010	0.039	0.102

The Birthrate Functions

We now present, finally, expressions for the birthrate functions Φ^{pi}_{mknl} and Φ^{pi}_{mk}, which are defined such that

$$\Phi^{pi}_{mknl}\; d\bar{X}_{pi}\, d\bar{u}_{pi} dt$$

is the number of particles created in the time dt in an element $d\bar{x}_{pi}d\bar{u}_{pi}$ of phase space as the result of collision of particles mk and nl, and

$$\Psi^{pi}_{mk}\; d\bar{x}_{pi} d\bar{u}_{pi} dt$$

is the mathematical expectation of the number of particles created in the same element and in the same period of time due to spontaneous disintegrations.

The total number of particles created by the first route will be

$$\Phi^{pi}\,(\bar{x}_{pi},\bar{u}_{pi},t)\; d\bar{x}_{pi} d\bar{u}_{pi} dt = \frac{1}{2}\, d\bar{x}_{pi} d\bar{u}_{pi} dt \sum_{mk}\sum_{nl} \iint_{-\infty}^{\infty} E_{mknl} T^{pi}_{mknl} d\bar{u}_{mk} d\bar{u}_{nl}\;, \tag{A.79}$$

and by the second

$$\Psi^{pi}_{mk}\,(\bar{u}_{mk},\bar{u}_{pi})\; d\bar{x}_{pi} d\bar{u}_{pi} dt = d\bar{x}_{pi} d\bar{u}_{pi} dt \iint_{-\infty}^{\infty} f_{mk} N^{pi}_{mk}\, d\bar{u}_{mk} d\bar{x}_{mk}\;. \tag{A.80}$$

This concludes our analysis of the auxiliary expressions; we are now in a position to derive the system of integral equations.

The System of Integral Equations as Postulated by Vallander

The results presented in this section were obtained on the basis of the works of the Vallander school (1960-1964) but are not exactly the same.

About the point $\bar{x}_p$ [which may be either in a gaseous volume (p = 3) or on a surface element (p = 0, 1, 2)] we set apart an element of volume $d\bar{x}_p$ and calculate the number of particles with velocity between $\bar{u}_{pi}$ and $\bar{u}_{pi} + d\bar{u}_{pi}$ that will be contained in that volume at the instant t. In other words, we are interested in the quantity

$$dn^{pi} = f_{pi}\,(\bar{x}_{pi},\bar{u}_{pi},t)\; d\bar{x}_{pi} d\bar{u}_{pi}\;.$$

It is understood from our earlier remarks that the number dn^{pi} will include dn_3^{pi} particles created in the gas both as the result of collisions (dn_3^{0pi}) and in spontaneous production (dn_3^{1pi}):

$$dn_3^{pi} = dn_3^{0pi} + dn_3^{1pi}\;. \tag{A.81}$$

The other components will be particles created by collision and spontaneously on different surface elements:

$$dn_2^{pi} = dn_2^{opi} + dn_2^{1pi} , \tag{A.82}$$

$$dn_1^{pi} = dn_1^{opi} + dn_1^{1pi} , \tag{A.83}$$

$$dn_0^{pi} = dn_0^{opi} + dn_0^{1pi} . \tag{A.84}$$

All of these particles provide a corresponding contribution if they can traverse the path from their birth-site to $d\bar{x}_p$ (i.e., without disintegrating and without colliding).

$$dn^{pi} = \sum_{k=0,1} \sum_{l=0}^{3} dn_l^{kpi} . \tag{A.85}$$

Here k is an index denoting the mode of production of particles of type i, l is an index denoting the position of the source, p indicates the position of the reaction products.

We now have to derive expressions for the distribution functions f_{pi} (i.e., f_{3i}, f_{2i}, f_{1i}, f_{0i}), expressed in terms of the functions derived earlier.

Let us begin with the expression for f_{3i}.

Let (see Fig. A.5) the elementary volume $d\bar{x}$ be a right circular cylinder with bases dS and generatrices dH parallel to the vector $\bar{u}$.

From the point $\bar{x}$ we draw a line opposite the direction of the vector $\bar{u}$ until it meets the surface at a point $\bar{x}_S$. If the ray does not meet the surface, we assume that $\bar{x}_S$ is at infinity.

On the line $\overline{xx_S}$ we pick out the point $\bar{x}_\tau$ and construct around it a slab such that its base is parallel to the base of dS and its generatrices are the same as those of dx, but the height dh is of a much smaller order than dH.

Let

$$\bar{x}_\tau = \bar{x} - (t-\tau)\bar{u} , \tag{A.86}$$

then the particles created in the slab about $\bar{x}_\tau$ in the time τ to $(\tau + d\tau)$ (where $dt = dH/|\bar{u}|$ is the residence time of particles with velocity $\bar{u}$ in the element dH) and not subjected to collision or disintegration will enter $d\bar{x}$ at the time t.

We call τ_S the time at which the points $\bar{x}_S$ most distant from $\bar{x}$ must emit particles with a velocity $\bar{u}$ in order for them to be within the element $d\bar{x}$ at the time t.

The number dn_3^{3i} [see Eq. (A.81)] is given by the equation

$$dn_3^{3i} = d\bar{x}d\bar{u} \int_{\tau_S}^{t} N_{3i}(\bar{x},\bar{u},\tau,t) [\Phi_{3k3l}^{3i}(\bar{x} - (t-\tau)\bar{u},\bar{u},\tau') + \Psi_{3k}^{3i}(\bar{x} - (t-\tau)\bar{u},\bar{u},\tau)]\, d\tau . \tag{A.87}$$

As already mentioned, the number dn^{3i} of particles with which we are concerned includes those created (spontaneously dn_l^{13i} and in collision dn_l^{03i}) on different surface elements.

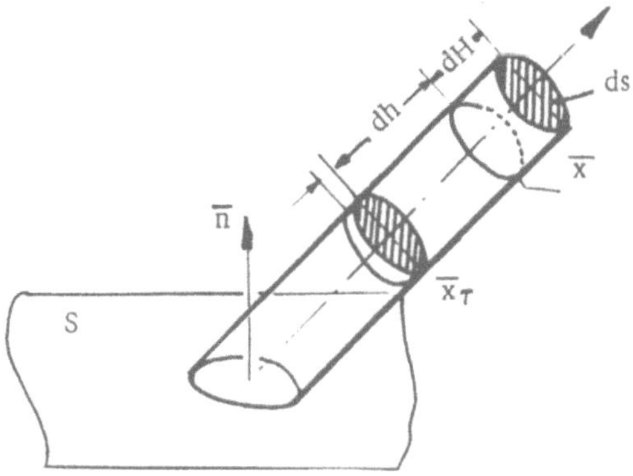

Fig. A.5. Calculation of creation functions distributed continuously throughout space (Vallander).

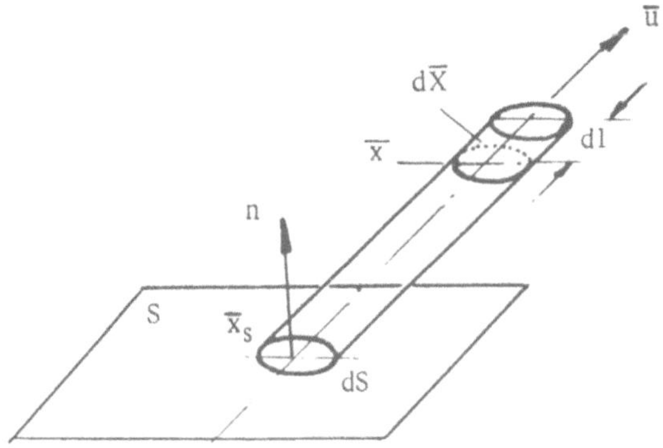

Fig. A.6. Calculation of the boundary creation functions (Vallander).

We assume that the steps have a density

$$\phi_L = \frac{dL}{dS}, \tag{A.88}$$

where dL is a linear element of the steps, dS is an element of area, and the creation of the remaining particles adsorbed in the surface layer of the steps and on the surface of the crystal is uniformly distributed over the length of the steps and over the surface.

To carry out these calculations we make use of Fig. A.6. As the figure clearly indicates, S is the surface on which particles impinge from the gaseous phase and which in turn is a source of ejected and spontaneously emitted particles, $\overline{x}$ is the position of the element $d\overline{x}$, which, unlike the $d\overline{x}$ shown in Fig. A.5, is a skewed cylinder. Its base is parallel to dS, and its generatrices are parallel to the vector $\overline{u}$. Everything that is created in one way or another in dS enters the element $d\overline{x}$ as long as it is not obstructed in its path.

During the time from τ_S to $\tau_S + dt$ the surface element dS emits particles in the pertinent velocity interval from u to $\overline{u} + d\overline{u}$, the number of such particles being proportional to dx:

$$d\overline{x} = |\overline{u}_n|\, dSdt, \tag{A.89}$$

where $\overline{u}_n$ is the projection of the velocity $\overline{u}$ on the normal $\overline{n}$ to the surface.

Using Eqs. (A.88), and (A.89), as well as the definitions of the creation functions and probability N_{3i} of an uneventful existence, we obtain

$$dn_2^{3i} + dn_1^{3i} + dn_0^{3i} = \frac{d\overline{x}d\overline{u}}{|\overline{u}_{ni}|} N_{3i}(\overline{x},\overline{u},\tau_s,t)\,[\Phi^{3i}_{3k2l} + \phi_L\Phi^{3i}_{3k1l} + \Phi^{3i}_{3kol} + \Phi^{3i}_{2k2l} + \phi_L\Phi^{3i}_{2k1l} +$$
$$+ \Phi^{3i}_{2kol} + \phi_L\Phi^{3i}_{1k1l} + \Phi^{3i}_{1kol} + \Phi^{3i}_{okol} + \Psi^{3i}_{2l} + \phi_L\Psi^{3i}_{1l} + \Psi^{3i}_{ol}]\,. \tag{A.90}$$

Using Eqs. (A.85), (A.87), (A.90), we obtain

$$f_{3i} = \frac{N_{3i}}{|\bar{u}_{n3i}|}\left(\sum_{m=0}^{3}\sum_{n=0}^{2}\Phi^{3i}_{mknl} + \sum_{m=0}^{m=2}\Psi^{3i}_{nl}\right) + \int_{\tau_S}^{t} N_{3i}\,(\Phi^{3i}_{3k3l} + \Psi^{3i}_{3k})\,d\tau, \tag{A.91}$$

where the terms containing $1l$ are multiplied by ϕL. Analogous formulas can be derived for the rest of the f_{pi}. However, we shall merely present the final equations:

$$f_{2i} = \frac{N_{2i}}{|\bar{u}_{n2i}|}\left(\sum_{m=0}^{3}\Phi^{2i}_{mk1l} + \Psi^{2i}_{1l}\right) + \int_{\tau_S}^{t} N_{2i}\left[\sum_{m=0}^{3}(\Phi^{2i}_{mkol} + \Phi^{2i}_{mk1l}) + \Psi^{2i}_{ok} + \Psi^{2i}_{2k}\right] d\tau \tag{A.92}$$

$$f_{1i} = \int_{\tau_S}^{t} N_{1i}\left[\sum_{m=0}^{3}\sum_{n=0}^{3}\Phi^{1i}_{mknl} + \sum_{m=0}^{2}\Psi^{1i}_{mk}\right] d\tau . \tag{A.93}$$

We may regard the beginnings of the steps and the shoulders to the steps as their boundaries. We have neglected this in Eq. (A.93), since allowance for the boundary at this point would require one more condition (m, n, p = 4):

$$f_{oi} = \int_{\tau_S}^{t} N_{oi}\left[\sum_{m=0}^{3}\sum_{n=0}^{3}\Phi^{oi}_{mknl} + \sum_{n=0}^{3}\Psi^{oi}_{nl}\right] d\tau , \tag{A.94}$$

$$f_{pi} = \frac{N_{pi}}{|\bar{u}_{npi}|}[\Phi^{pi}_{loc} + \Psi^{pi}_{loc}] + \int_{\tau_S}^{t} N_{pi}\,(\Phi^{pi}_{dis} + \Psi^{pi}_{dis})\,d\tau , \tag{A.95}$$

where Φ^{pi}_{loc} and Ψ^{pi}_{loc} reflect local sources of particles of type pi as the result of particle interaction and spontaneous process, respectively, Φ^{pi}_{dis} and Ψ^{pi}_{dis} are the same sources but distributed over the surface or throughout the volume, etc.

In Eqs. (A.91)-(A.94) all the functions f_{pi} are expressed in terms of the quantities Φ^{pi}_{mknl} and Ψ^{pi}_{mk}, as well as N_{pi}, which in turn is also related to f_{pi} by means of σ_{mknl}, T^{pi}_{mknl}. The latter quantities must be determined, for example, by experiment or quantum mechanical calculations, and they must be known.

Then equations of the type (A.95) may be written in the form of an integral equation of the function f_{pi}:

$$f_{pi} = V f_{pi} \tag{A.96}$$

where V is an integral operator. It is shown in the works of Vallander and Belova (A1, 7 and 45) that when the operator

$$\frac{\partial}{\partial t} + u_1\frac{\partial}{\partial x_1} + u_2\frac{\partial}{\partial x_2} + u_3\frac{\partial}{\partial x_3} + g_1\frac{\partial}{\partial u_1} + g_2\frac{\partial}{\partial u_2} + g_3\frac{\partial}{\partial u_3}$$

is applied to Eq. (A.96), we obtain the kinetic equations of Boltzmann, generalized to the case of gases of variable composition.

It has long ago become a tradition to mention the relationship between the Boltzmann integrodifferential kinetic equation (see, e.g., de Groot and Mazur, 1962; MacDonald, 1962) and the differential equation of Fokker and Planck, the diffusion (thermal conduction) equation, and the chemical reaction kinetic equation. We will not dwell any further on this problem.

In light of the large number of functions and coefficients that have been brought to bear on the problem, the relations derived are better thought of as a plan for investigation, rather than the final results. This relates to the fact that many of the functions and coefficients still remain unknown.

In the next section we will exhibit somewhat more meaningful results and implications of the equations derived above.

Adsorption Equations

In certain portions of this section we will have occasion to draw upon the works of Filippov (A1, 162, A2, 272).

Let $A^i(\bar{x}, t)$ be the mathematical expectation of the number of type i particles existing at the time t in the adsorbed state on a unit area at the point $\bar{x}$.

It is plain to see that $A^i(\bar{x}, t)$ may be expressed either in terms of the distribution functions

$$A^i(\bar{x},t) = \int_{-\infty}^{\infty} f_{2i}\, d\bar{u}_{2i} + \int_{-\infty}^{\infty} f_{1i}\, d\bar{u}_{1i}\,, \tag{A.97}$$

or else

$$A^i(\bar{x},t) = A_1^i + A_2^i\,, \tag{A.98}$$

where A_1 is the mathematical expectation of the number of particles leaving a unit area from the time t_0 to the time of observation t, while A_2 is the mathematical expectation of the number of particles arriving at the same place in the time interval $t_0 - t$ but without departing before the time of observation t.

We assume that we know A_0^i, the number of particles per unit surface at the time t_0, and that the probability remains the same from t_0 until t (i.e., neither spontaneous nor induced desorption take place, nor is there chemical reaction or diffusion, etc.) and equal to $\tilde{\alpha}^i(\bar{x}, T, f_{1i}, f_{2i}, A^i, t_0, t)$, whereupon

$$A_1^i = \tilde{\alpha}^i A_0^i\,. \tag{A.99}$$

A particle contributes to the mathematical expectation A_2^i if the following chain of events has taken place: 1) The particle appeared on the surface either from the ambient gaseous sphere or as the result of diffusion on the surface or of chemical reaction; 2) remaining on the clean or occupied surface, the particle was adsorbed thereon; 3) during the time interval from the instant of adsorption t_0 until the time of observation t, the particle remained on the surface:

$$A_2^i = \tilde{\alpha}^i \int_{t_0}^{t} \tilde{K}^i\, dt\,, \tag{A.100}$$

where $\tilde{K}^i$ is the probability of the complex event 1), 2).

Bearing Eq. (A.97) in mind, we can express $\tilde{\alpha}^i$ and $\tilde{K}^i$ in terms of the creation functions. It is in our best interest to relate $\tilde{\alpha}^i$ to more fundamental concepts than Φ^{pi}_{mknl} and Ψ^{pi}_{mk}.

Let us assume, as is usually done (see, e.g., Sec. 2), that the particle is desorbed if its energy exceeds the binding energy $Э^{pi}$ (where p is an index of the type of bond).

The probability of spontaneous desorption during a certain time interval coincides with the probability that the particle energy will at least just exceed the binding energy. Let the probability of this event during the time dt be $N(Э^{pi}, T, \bar{x})dt$, where T is the surface temperature.

To determine the surface temperature it is necessary to solve the problem of the thermal dissipation of energy imparted by the incident flows. We will assume here that this problem has been solved and that the temperature is already known.

The particles adsorbed on the surface may be treated entirely under quasiequilibrium conditions if the characteristic time of variation of T is greater than the vibrational period of the adsorbed particles (which is $\sim 10^{-13}$ sec). Then the distribution functions f_{2i}, f_{1i}, f_{0i} may be regarded to a first approximation as equilibrium distributions.

In view of the fact that desorption is a discontinuous process, the probability $\tilde{\alpha}_0^i$ that desorption will not occur spontaneously during the period from τ to t is equal to

$$\tilde{\alpha}_0^i = \exp\left[-\int_\tau^t N(Э^{pi}, T, \bar{x})\, dt\right]. \tag{A.101}$$

Let the particle have different types of bond with the surface, for example, van der Waals, covalent, ionic, etc., the weights of which are G_1, G_2, . . ., etc., so that the probability of spontaneous desorption, averaged over all possible bonds, is expressed as

$$\tilde{\alpha}_0^{pi} = \exp\left[-\int_\tau^t [G_1 N_1 + G_2 N_2 + \cdots]\, dt\right]. \tag{A.102}$$

In order to assess the role of desorption in collision between particles of external flux, we write the probability of the particle energy state γ_p:

$$\frac{\mathcal{G}_{\gamma_p} \cdot \gamma_p \cdot \exp(-\gamma_p/kT)}{\sum\limits_{\gamma_p} \mathcal{G}_{\gamma_p} \exp(-\gamma_p/kT)}, \quad p = \text{const} \tag{A.103}$$

where $\mathcal{G}_{\gamma_p}$ is the statistical weight of the state with energy γ_p.

Let $\mathcal{P} = \mathcal{P}(\bar{x}, T, A^i, \bar{u}_l, l)$ be the probability, averaged over all bond types and energy states, that a particle adsorbed in the region of $\bar{x}$ will be desorbed as the result of collision with a given particle l.

$\mathcal{P}$ is defined by the formula

$$\mathcal{P}(\bar{x}, T, A^i, \bar{u}_l, l) = \frac{G_p(\bar{x}) \sum\limits_p \sum\limits_{\gamma_p} \mathcal{G}_{\gamma_p} \exp(-\gamma_p/kT) P_p(\bar{x}, \gamma_p, \bar{u}_l, l)}{\sum\limits_p \sum\limits_{\gamma_p} \mathcal{G}_{\gamma_p} \gamma_p \exp(-\gamma_p/kT)}. \tag{A.104}$$

where $P_p = P_p(\bar{x}, \gamma_p, \bar{u}_l, l)$ is the probability of desorption of a particle bound to the surface $\bar{x}$ by a type p bond in the energy state γ_p during impact with a particle of type l.

Hence the probability that desorption will not occur due to collision with particles, equal to

$$\tilde{\alpha}_1^i = \frac{\tilde{\alpha}^i}{\tilde{\alpha}_0^i}, \tag{A.105}$$

is determined from the equation

$$\tilde{\alpha}_1^i = \exp\left[-\int_{\tau}^{t}\int_{u_n<0} |\bar{u}_n| \sum_l \mathscr{P}(\bar{x},T,A^i,\bar{u}_l,l)\cdot f_{3l}(\bar{x},\bar{u}_l,\tau)\,\sigma_l(\bar{u}_l)\,d\bar{u}_l dt\right]. \tag{A.106}$$

The above calculations were carried out on the assumption that the principal contributors to desorption are particles arriving from the gaseous volume, for which the velocity distribution function $f_{3l}(\bar{x}, \bar{u}, \tau)\sigma_l(\bar{u}_l)$ may be expressed in terms of f_{2j}, f_{1j}, f_{0j} with cross section of the type $\sigma_{ml,nj}$. Here σ_l is the collision cross section of a traveling particle of type l with one of the surface particles.

We now give an interpretation of the function $\tilde{K}^i$ appearing in Eq. (A.100), with certain simplifying assumptions and notations:

1) The fraction of the surface occupied by particles of type i is equal to $A^i(x, t)\sigma$, where σ is the fraction of the surface ascribed to one particle such that the surface is completely covered.

2) The probability of adsorption on the free surface is $\sum_{p=0}^{2} \Phi_{3loj}^{pi}$, on the occupied surface $\sum_{p=0}^{2}\sum_{n=0}^{2} \Phi_{3lnj}^{pi}$.

3) The diffusion probability is negligibly small.

It must be understood that if we were to assume that no chemical reactions occurred on the surface, the probability of adsorption on the free surface would be described as $\sum_{p=0}^{2} \Phi_{3ioi}^{pi}$ and on the occupied surface as $\sum_{p=0}^{2}\sum_{n=0}^{2} \Phi_{3ini}^{pi}$.

If a contribution to the number of adsorbed particles is made by particles of several types, it is necessary to add these contributions.

Then

$$\tilde{K}^i = \left[(1 - A^i\sigma)\sum_{p=0}^{2} \Phi_{3ioi}^{pi} + A^i\sigma \sum_{p=0}^{2}\sum_{n=0}^{2} \Phi_{3ini}^{pi}\right] \times \int_{(u_n<0)} |\bar{u}_n|\, f_{3i}\, d\bar{u} dS. \tag{A.107}$$

As indicated earlier, $|\bar{u}_n| f_{3i}$ is the mathematical expectation of type i particles [from the velocity space $\bar{u}-(\bar{u}+d\bar{u})$] colliding with a unit area per unit time.

Substituting into (A.98), (A.99), and (A.100), we obtain

$$A^i = \tilde{\alpha}^i \left(A_0^i + \int_{t_0}^{t} \tilde{K}^i d\tau\right) \tag{A.108}$$

If in addition to the restrictions imposed ahead of Eq. (A.107) we assume that:

4) $\tilde{\alpha}^i$ and $\tilde{K}^i$ do not depend on A^i (this is clearly valid for small coverage); 5) the rate of change of f_{3i} is small in comparison with the rate of the adsorption processes; 6) all bonds with the surface are identical; 7) skipping effects are absent; then, as found by Filippov, Eq. (A.108) is transformed into the equation for the Langmuir adsorption isotherm:

$$A^i = \frac{\gamma_{\bar{x}}\, \sigma^{-1} P}{1 + \gamma_{\bar{x}} P}, \tag{A.109}$$

where P is the pressure, $\gamma_{\bar{x}}$ is related in simple fashion to the functions introduced earlier:

$$\gamma_{\bar{x}} = \frac{1}{P} \sum_{p=0}^{2} \Phi_{3i0i}^{pi} \int_{(u_N<0)} |\bar{u}_n|\, f(\bar{u})\, d\bar{u}, \tag{A.110}$$

where $f(\bar{u})$ is the equilibrium Maxwell distribution function.

If f is not quite an equilibrium function but depends on the time, then as a first simplification we will assume that f depends only on the interval of observation:

$$f_i = f_i(\bar{x}, \bar{u}, t - t_0), \tag{A.111}$$

and the solution (A.108) assumes the form

$$A^i = \frac{\gamma_{\bar{x}}\, \sigma^{-1} P}{1 + \gamma_{\bar{x}} P} + \left(A_0^i - \frac{\gamma_{\bar{x}}\, \sigma^{-1} P}{1 + \gamma_{\bar{x}} P}\right) e^{-(\gamma_{\bar{x}} \sigma P + N)t}, \tag{A.112}$$

where N is the same as in (A.101).

It is clear that Eq. (A.112) with $t \to \infty$ leads to the Langmuir equation (A.109).

Systems of Integral Equations for Molecular Flows

In this section we give a system of integral equations differing from those of the type (A.30), (A.32)-(A.36), first, in that the processes which they describe are not stationary; second, in that they do not allow for the possible existence of different chemical compounds capable of being converted one into the other. In addition, the first of these systems (A.113) refers to monoenergetic flows, although they can readily be transformed into a system of equations for total flows.

We will use the same notation here as in the derivation of Eqs. (A.1)-(A.6), but we will augment it with superscripts indicating the type of particle. For example, the symbol I_r^i indicates the monokinetic flux of type i molecules incident on a unit surface of type r per unit time. The superscripts take on the range of values $i = 1, 2, \ldots, i_{max}$, while the subscripts have the range $r = 1, 2, \ldots, r_{max}$. The indices are arguments of the functions to which they are attached. The indexing of the creation functions is the same as before.

Our cumulative experience of the preceding pages in the derivation of integral equations of various types permits us to construct the system of integral equations for the effective molecular flux without derivation. Analogous systems could also be written down for other classes of molecular fluxes [natural emission, the absorbed part, reflected flux, incident flux, resultant adsorption – see the text associated with Eqs. (A.1)-(A.6). We could also write down the system of algebraic equations. The reader may wish to do these operations as an exercise.

$$E_e^i(\bar{x},t) = E_c^i(\bar{x},t) + \int_{(S)} F^i(\bar{x},\bar{x}_1) \Big[\overbrace{\rho^i(\bar{x})\, E_e^i\Big(\bar{x}_1, t - \frac{|\bar{x}-\bar{x}_1|}{u^i}\Big)}^{\text{I}} + \overbrace{E_e^k\Big(\bar{x}_1, t - \frac{\bar{x}-\bar{x}_1}{u^k}\Big)\Psi_k^i}^{\text{II}} + \overbrace{E_e^l\Big(\bar{x}_1, t - \frac{\bar{x}-\bar{x}_1}{u^l}\Big) E_e^m\Big(\bar{x}_1, t - \frac{\bar{x}-\bar{x}_1}{u^m}\Big)\Phi_{lm}^i}^{\text{III}} \Big] d\bar{x}_1, \tag{A.113}$$

where the Roman numeral I refers to ordinary reflection of particles without chemical conversion, II to the components associated with processes of spontaneous production of particles arriving at $\bar{x}$ from $\bar{x}_1$, III to processes of production as the result of collision at the point $\bar{x}$ between particles of type l and type m arriving from $\bar{x}_1$. In the event that reactions are possible by several routes, it becomes necessary to utilize appropriate summations.

The kernels of the equations, $F^i(\bar{x}, \bar{x}_1)$ include the probability of uneventful transfer from the point $\bar{x}$. It is quite obvious that as $u^i \to \infty$ the equations cease to determine future events.

A more general criterion is

$$\frac{\bar{x}-\bar{x}_1}{u^i} \ll t. \tag{A.114}$$

For

$$\frac{\bar{x}-\bar{x}_1}{u^i} \approx t \tag{A.115}$$

the equation becomes quasistationary.

We conclude by writing out the system of integral equations for the effective molecular flux when the surface of the reservoir can be divided into separate zones (see the section "Statement of the Problem"), which act as sources and sinks of different types of particles.

$$E_{er}^i(\bar{x},\bar{u},t) = E_{cr}^i(\bar{x},\bar{u},t) + \sum_{m^{min}}^{m^{max}} \sum_{k=0}^{k^{max}} \int_{x_m^{min}}^{x_m^{max}} F(\bar{x},\bar{x}_m)\,\alpha_{ik} E_e^k\Big(\bar{x}_m, \bar{u}^k, t - \frac{\bar{x}-\bar{x}_m}{\bar{u}^k}\Big) d\bar{x}_m, \tag{A.116}$$

where α_{ik} is the probability of obtaining a particle of type i from those of type k arriving at $\bar{x}$ (α_{ik} can be expressed in terms of Φ_{mknl}^{pi}, Ψ_{mk}^{pi}, f_{pi}).

In particular, α_{ii} is the ordinary reflection coefficient for type i particles arriving from $\bar{x}_1$ at $\bar{x}$.

In order to take into account bimolecular reactions which can elicit the emission of type i particles from the vicinity of the point $\bar{x}$, it is necessary to bring in one more component:

$$\sum_{m_1^{min}}^{m_1^{max}} \sum_{m_2^{min}}^{m_2^{max}} \sum_{k_1=0}^{k_1^{max}} \sum_{k_2=0}^{k_2^{max}} \int_{x_{m_1}^{min}}^{x_{m_1}^{max}} \int_{x_{m_2}^{min}}^{x_{m_2}^{max}} F(x,x_{m_1}) F(x,x_{m_2}) \alpha^i_{k_1k_2} \times$$

$$\times E_e^{k_1}\left(\bar{x}_{m_1}, \bar{u}^{k_1}_{m_1}, t - \frac{|\bar{x} - \bar{x}_{m_1}|}{u^{k_1}}\right) E_e^{k_2}\left(\bar{x}_{m_2}, \bar{u}^{k_2}_{m_1}, t - \frac{|\bar{x} - \bar{x}_{m_2}|}{u^{k_2}}\right) d\bar{x}_{m_1} d\bar{x}_{m_2},$$

where $\alpha^i_{k_1k_2}$ is the probability of obtaining type i particles when the collision of type k_1 and type k_2 particles is a certainty.

In this way we readily account for the possibility of even more complex reactions (as the result of, for example, three-way collisions).

If by analogy with (A.116) we write down the equations for the remaining classes of molecular flow (A.1)-(A.6), we essentially arrive at the answer to the questions set forth in the "Statement of the Problem."

It is hoped that the reader can now carry this through as an exercise.

SUMMARY OF NOTATION

a evaporation or desorption rate, particles/cm^2 sec
a_1, a_2 evaporation rates of monomers and dimers from bottom of cylinder, particles/cm^2 sec
a_k Knudsen thermal accommodation coefficient
a_k^r, a_k^v, a_k^t accommodation coefficients for rotational, vibrational, and translational energy
a_m Langmuir evaporation rate, g/cm^2 sec
a_{me} effusion rate from Knudsen cell, g/cm^2 sec
a_l lattice constant
a_e effusion rate from Knudsen cell, particles/cm^2 sec
A number of adsorbed particles
b_1, b_2 evaporation rates of monomers and dimers from wall, particles/cm^2 sec
B mean roughness height
c_i concentration of component i
d density of gas at unit pressure
d_p density of catalyst grain
D diffusion coefficient
E number of particles emitted per cm^2 sec
f_m Maxwell momentum-transfer coefficient
$F(|z-z_1|)$ or $F(z, z_1)$ kernel of integral equation
h Planck constant
I number of particles moving in unit solid angle
I_0 the same, along normal to surface
J total flux of particles through tube
k reaction rate constant
k_B Boltzmann constant
κ coefficient of volume absorption of light
l dimensionless length of cylinder
L length of a cylinder, diaphragm, capillary, etc.
m mass of particle, g
M molecular weight
n density of gas, particles/cm^3
$n_D^{(z)}$ number of dimers incident at point z of cylinder wall, cm^{-2}sec^{-1}
$n_M^{(z)}$ number of monomers incident at point z of cylinder wall, cm^{-2}sec^{-1}
n_s equilibrium density of gas, particles/cm^3
N Avogadro's number
p pressure
p_O pressure at aperture of Knudsen cell
p_T pressure at the condensed phase
q effective evaporation coefficient
Q heat of adsorption, activation, evaporation, etc.
r radius of capillary or cylinder
r_c^{AB} partial reflecting power
R universal gas constant, or, when specially stated, radius
R_0 speed of reflection
s perimeter of cross section of tube

S area of evaporation surface

S_0 area of aperture of Knudsen cell

S_T area of condensed phase

S_x are of macrosurface of catalyst

T temperature, °K

u speed of particle

u_0 rate of slip

U rate of reaction

V volume of system

V_g volume of pores in one gram of catalyst

V total volume of catalyst grain

V_r total volume of reactor

$w_{dd}(x)$ probability of a particle traveling without collision from a disk to another disk separated from it by a distance x

$w_{db}(x)$ the same, but from a disk to a band of width dx

$w_{bd}(x)$ the same, but from the band of width dx to the disk

$w_{bb}(x)$ the same, but from band to band

$w(\rho')$ probability of traveling a distance ρ' without collision

W Clausing coefficient of a cylinder

$\vec{W}$ Clausing coefficient for a convergent cone (or for a vessel in the direction indicated by the arrow)

$\overleftarrow{W}$ Clausing coefficient for a divergent cone

W_α Clausing coefficient for tubes whose walls have conuensation coefficient α

W_0 Clausing coefficient of aperture of Knudsen cell

W_ϑ Clausing coefficient of a pipe bend which turns through the angle ϑ

α condensation (or evaporation) coefficient for plane surface

β coefficient of evaporation from walls

δ_a δ_b evaporation coefficients of dimers from bottom and wall of cylinder

ε emittance

η viscosity coefficient

ϑ angle of incidence for reflection of beams

Θ characteristic temperature of a solid

$\varkappa$ effective depth of reaction zone (Section 11)

$\varkappa'$ degree of dissociation of dimers into monomers

λ mean free path

λ_B de Broglie wavelength

μ effusion rate from a cylinder, particles/cm^{2}sec

$\bar{\mu}$ effusion rate from a cylinder, if the number of impacts of particles per cm^{2}sec at the saturated vapor pressure is normalized to unity.

μ_D dimer effusion rate, cm^{-2}sec^{-1}

μ_D^b rate of departure of dimers from bottom of cylinder, cm^{-2}sec^{-1}

μ_M^b rate of departure of monomers from bottom of cylinder, cm^{-2}sec^{-1}

$\bar{\mu}_D$, $\bar{\mu}_D^b$, $\bar{\mu}_M^b$ the same, when the rate of departure of particles per cm^{2}sec under saturation conditions is normalized to unity

ν frequency of vibration of an atom about its position of equilibrium

ρ reflection coefficient

$\rho^\perp$, ρ^B reflection coefficients of particles at normal incidence to surface or from the direction of the point B

ρ_{aDD}, ρ_{bDD} coefficients of reflection of dimers from bottom and walls of cylinder

ρ'_{aDM}, ρ'_{bDM} fractions of dimers reflected from bottom and walls in form of two monomers

ρ''_{aDM}, ρ''_{bDM} fractions of dimers reflected from bottom and walls in form of a single monomer
ρ_M monomer reflection coefficient
σ, τ pumping times (Section 5)
τ generally time or period (with various indices), but in Section 10, coefficients of condensation of trimers accompanied by chemical reactions
φ glancing angle
$\Phi(z)$ lengthwise distribution of number of particles/cm^{2}sec in a cylinder
Ψ used to denote angles, but in Section 11, porosities
ω', ω'' resistances of tube and aperture to molecular flow
$\bar{\omega}(\alpha)$ value, averaged over the entire surface of an evaporating substance (with evaporation coefficient α), of the probability of escape of a particle from the surface into vacuum
$\bar{\omega}_p(\alpha)$ the same as $\bar{\omega}(\alpha)$, but for the special case of a surface covered with grooves of triangular cross section, with vertex angle ϑ
Ω solid angle
Э energy (Section 1)
P roughness (Section 11)

Note. Not included here are a number of symbols used in the figures, in intermediate calculations, and in the appendix. The meanings of these notations are clear from the context.

PUBLISHER'S NOTE

The following Soviet journals cited in this book are available in cover-to-cover translation:

Russian Title	English Title	Publisher
Zhurnal Tekhnicheskoi Fiziki	Soviet Physical-Technical Physics	American Institute of Physics
Doklady Akademii Nauk SSSR	Soviet Physics-Doklady	American Institute of Physics
Zhurnal Fizicheskoi Khimii	Russian Journal of Physical Chemistry	The Chemical Society (London)
Zhurnal Éksperimental'noi i Teoreticheskoi Fiziki	Soviet Physics-JETP	American Institute of Physics
Zhurnal Neorganicheskoi Khimii	Russian Journal of Inorganic Chemistry	The Chemical Society (London)
Uspekhi Matematicheskaya Nauk	Russian Mathematical Surveys	Cleaver-Hume Press, Ltd.

LITERATURE CITED

Adzumi, H., Bull. Chem. Soc. Japan 12:285 (1937).

Agrest, M. M., Zh. Tekhn. Fiz. 28:1340 (1958).

Agrest, M. M., M. Z. Maksinov, and A. K. Khmel'nitskii, Zh. Tekhn. Fiz. 28:1345 (1958).

Alty, T., Can. J. Res. 4:457 (1931); Phil. Mag. 15:82 (1933); Proc. Roy. Soc. A149:104 (1935); Proc. Roy. Soc. A161:68 (1937).

Anderson, R. B., and A. M. Whitehouse, Ind. Eng. Chem. 58:1011 (1961).

Balson, E. W., Surface Phenomena in Chemistry and Biology, Pergamon Press, Inc., New York (1958), p. 117; J. Phys. Chem. 65:1151 (1961).

Bartlett, A.C., Phil. Mag. 40:111 (1920).

Baule, B., Gasen Ann. Phys. 44:145 (1914).

Blokh, A. G., Fundamentals of Radiative Heat Transfer, Gosénergoizdat (1962).

Bouguer, P., Optical Treatise on the Gradation of Light, University of Toronto, Canada (1961) [Russian translation] A. A. Gershun (ed.), Izd. Akad. Nauk SSSR (1950).

Brewer, L., and J. S. J. Kane, Phys. Chem. 59:105 (1955).

Buckley, H., Phil. Mag. 4:753 (1927); Phil. Mag. 6:447 (1928); Recueil Trav. Com. Internat. d'Eclair, 7 Sess.: 888 (1928); Phil. Mag. 17:576 (1934).

Carberry, J. J., J. Am. Inst. Chem. Eng. 6:460 (1960).

Chambré, P. L., and A. Acrivos, J. Appl. Phys. 27:1322 (1956).

Chambré, P. L., J. Chem. Phys. 32:24 (1960).

Chapman S., and T. G. Cowling, The Mathematical Theory of Nonuniform Gases, Cambridge University, Press, New York (1952).

Cherepnin, N. V. (ed.), Modern Vacuum Technique [Russian translation] IL (1963).

Chernov, A. A., Dokl. Akad. Nauk SSSR 117 (6):983 (1957).

Clausing, P., Verslag. Amsterdam 35:1023 (1926); Leidener dissert. Amsterdam (1928); Thesis Leyden (1929). Physica 9:65 (1929); Ann. Phys. 4:533 (1930); Ann. Phys. 4:567 (1930); Ann. Phys. 7:489 (1930); Ann. Phys. 7:569 (1930); Z. Phys. 66:471 (1930); Ann. Phys. 12:961 (1932).

Davis, D. H., J. Appl. Phys. 31:1169 (1960).

Davis, D. H., L. L. Levenson, and N. Milleron, Rarefied Gas Dynamics, Academic Press Inc., New York (1961).

de Boer, J. H., Dynamical Character of Adsorption, Oxford University Press, New York (1953).

de Broglie, L., Phil. Mag. 47:446 (1924).

de Groot, S. R., and P. Mazur, Non-Equilibrium Thermodynamics, North Holland Publishing Co., Amsterdam (1962).

Detkov, S. P., Zh. Fiz. Khim. 31:2367 (1957); Zh. Tekhn. Fiz. 30 (1):96 1956; Zh. Fiz. Khim. 34 (1):96 (1960).

Devienne, F. M., Flow and Heat Transfer in Rarefied Gases [Russian translation] IL (1962).

Devienne, F. M., J. Phys. Radium 17:257 (1953).

Devienne, F. M. (ed.), Rarefied Gas Dynamics (Proceedings of First Symposium, Nice), Pergamon Press, Inc., New York (1960).

Devonshire, A. F., Proc. Roy Soc. A158:158, 269 (1937).

De Vos, J. S., Physica 20:669 (1954).

Dushman, S., Scientific Foundations of Vacuum Technique, John Wiley & Sons, Inc., New York (1949).

Dushman, S., Production and Measurement of High Vacuum (1922); International Critical Tables, Vol. 1, (1926), p. 91; J. Franklin Inst. 211:689 (1931).

Emmett, P. H., Advan. Catalysis 1:65 (1948).
Esterman, I., and O. Stern, Z. Phys. 61:95 (1930).
Filimonov, B. V., Dokl. Akad. Nauk SSSR 150 (2):290 (1963).
Firsova, L. P., Zh. Fiz. Khim. 36:1322 (1962); Zh. Fiz. Khim. 36:1607 (1962).
Fok, V. A., Tr. Gos. Optich. Inst. 3 (28) (1924); Mat. Sb. 14(56):1-2 (1944).
Foote, P. D., Bull. Bur. St. 12:585 (1916) (cited by Walsh, 1919-1920, which see).
Fowler, R., Proc. Roy. Soc. A150:456 (1935) (cited by Massey and Burhop, 1958, which see).
Fox, J. W., A. C. U. Smith, Proc. Phys. Soc. 73:533 (1959).
Fraser, R. G. J., Molecular Rays, Cambridge University Press, New York (1931); Molecular Beams (Methuen's Monographs on Physics), Methuen, London (1937).
Frauenfielder, H., Helv. Phys. Acta 23:347 (1950).
Freeman, M. P., and K. F. Kolb, Phys. Chem. 67:217 (1963).
Frenkel', Ya. Zh., Zh. Eksperim. i Teor. Fiz. 16 (1):39 (1946).
Gaede, W., Ann. Phys. 41:289 (1913).
Geguzin, Ya. E., and N. N. Ovcharenko, Izv. Akad. Nauk SSSR, Otd. Tekhn. Nauk, Met. i Toplivo, No. 1:108 (1956).
German, O., Zh. Eksperim. i Teor. Fiz. 34(6):1470 (1958); Zh. Tekhn. Fiz. 32(9):1134 (1962).
Gershun, A. A., Tr. Gos. Optich. Inst. 9(87):23 (1933).
Giordmaine, J. A., and T. C. Wang, J. Appl. Phys. 31:463 (1960).
Gorokhov, L. N., Yu. S. Khodeev, and P. A. Akishin, Zh. Neorgan. Khim. 3:2597 (1958).
Groszkowski, J., High Vacuum Technology [Russian translation] IL (1957).
Günther, K. G.; Z. Angew. Phys. 9:550 (1957).
Hertz, H., Ann. Phys. 17:177 (1882) (cited by Knacke and Stranski, 1956).
Herzfeld, K. F., The Kinetic Theory of Matter [Russian translation] ONTI, NKTP SSSR (1935).
Hirschfelder, J. O., C. F. Curtiss, and R. B. Bird, Molecular Theory of Gases and Liquids, John Wiley & Sons, Inc., New York (1959).
Hirth, J., and G. Pound, J. Chem. Phys. 26:1216 (1957) [Russian translation] in: Lemmlein and Chernov (1959).
Holst, G., and P. Clausing, Physica 6:48 (1926); Physica 8:289 (1928).
Honigmann, B., The Growth and Form of Crystals [Russian translation] IL (1961).
Hougen, O. A., Ind. Eng. Chem. 53:509 (1961).
Hyde, E. P., Bull. Bur. St. 3:88 (1907) (cited by Walsh, 1919-1920, which see).
Iczkowski, R. P., J. L. Margrave, and S. M. Robinson, J. Phys. Chem. 67:229 (1963).
Inghram, M. G., and J. Drowart, International Symposium on High Temperature Technology, New York (1959).
Ivanov, B. S., and V. S. Troitskii, Zh. Tekhn. Fiz. 33(4):494 (1963).
Ivanovskii, A. I., and A. I. Repnev, Tr. Tsenti. Aerolog. Observ., No. 29:51 (1960).
Ivanovskii, A. I., and A. I. Repnev, Tr. Tsenti. Aerolog. Observ., No. 40:72 (1962).
Jaeckel, R., Kleinste Drucke, ihre Erzeugung und Messung (Very Low Pressures, Their Production and Measurement), Springer-Verlag, Berlin (1950).
Jepsen, D. W., and G. A. Somorjai, J. Chem. Phys. 39:1665 (1963).
Kennard, E. N. Kinetic Theory of Gases with an Introduction to Statistical Mechanics, New York (1938).
Knacke, O., and I. N. Stranski, Progress in Metal Physics (1956) [Russian translation] Usp. Fiz. Nauk 68:261 (1959).
Knauer, F., and O. Stern, Z. Phys. 53:781 (1929).
Knudsen, M., Ann. Phys. 28:75 (1909); Ann. Phys. 28:999 (1909); Ann. Phys. 29:179 (1909); Ann. Phys. 31:205 (1910); Ann. Phys. 31:633 (1910); Ann. Phys. 32:809 (1910); Ann. Phys. 33:1435 (1910); Ann. Phys. 34:593 (1911); Ann. Phys. 34:823 (1911); Ann. Phys. 35:389 (1911); Kongr. Solvey (1911); Abhandl. Dtsch. Buns. Ges. 3:116 (Halle, 1913); Ann. Phys. 44:525 (1914); Ann. Phys. 47:697 (1915); Ann. Phys. 48:1113 (1915); Ann. Phys. 83:797 (1927); Kinetic Theory of Gases, London, (1946); Kinetic Theory of Gases, third edition, London (1950).
Kolmogorov, A. N., Usp. Matem. Nauk, No. 5:5 (1938).

Kolmogorov, A. N., and N. A. Dmitriev, Dokl. Akad. Nauk SSSR 56(1):7 (1947).
Kolmogorov, A. N., and B. A. Sevast'yanov, Dokl. Akad. Nauk SSSR 56(8):783 (1947).
Korsunskii, M. N., and S. A. Vekshinskii, Zh. Eksperim. i Teor. Fiz. 15(10):593 (1945).
Ladenburg, R., and E. Thiele, Z. Phys. Chem. B7:161 (1930) (cited by Frazer, 1931, which see).
Landau, L. D., Phys. Z. Sowjetunion 8:489 (1935).
Landau, L. D., and E. M. Lifshits, Mechanics, Fizmatgiz (1958).
Langmuir, I., J. Am. Chem. Soc. 35:105 (1913); J. Am. Chem. Soc. 35:931 (1913); Phys. Rev. 14:1273 (1913) Phys. Rev. 8:149 (1916); Z. Elektrochem. 26:197 (1920).
Lemmlein, G. G., and A. A. Chernov (eds.), Elementary Processes of Crystal Growth [Russian translation] IL (1959).
Lennard-Jones, J. E., and A. F. Devonshire Proc. Roy. Soc. A156:6 (1936).
Littlewood, R., and E. Rideal, Trans. Faraday Soc. 52:1958 (1956).
Lozgachev, V. I., Zh. Fiz. Khim. 33 (12):2755 (1959); Zh. Fiz. Khim. 34(2):306 (1960); Izv. Akad. Nauk SSSR, Otd. Tekhn. Nauk, Met. i Toplivo, No. 4:34 (1961); Zh. Tekhn. Fiz. 32(8):1012 (1962); Zh. Tekhn. Fiz. 32 (8):1023 (1962); Zh. Tekhn. Fiz.. 32(9):1123 (1962); Zh. Fiz. Khim. 37:644 (1963).
Lyubitov, Yu. N., Proceedings of the Fifth Conference on the Physical and Chemical Foundations of Steel Production, Izd. Akad. Nauk SSSR (1961); Proceedings of the Sixth Conference on the Physical and Chemical Foundations of Steel Production, Izd. Akad. Nauk SSSR (1964); Inzh.-Fiz. Zh. 5(1):55 (1962); Proceedings of the Second Conference on the Technique of High-Temperature Research, Izd. Akad. Nauk (in press); Collection: Heat-Resistant Alloys, No. 8 Izd. Akad. Nauk SSSR (1962); Zh. Fiz. Khim. 37(8):1917 (1963).
MacDonald, D. K. C., Noise and Fluctuations; an Introduction, John Wiley & Sons, Inc., New York (1962).
Massey, H. S. W., and E. H. S. Burhop, Electronic and Ionic Impact Phenomena, The Clarendon Press, Oxford (1956).
Maxwell, C., Phil Trans. 170:231 (1880).
Mayer, H., Z. Phys. 52:235 (1928); Z. Phys. 58:373 (1929).
McKinley, J. D., and J. E. J. Vance, J. Chem. Phys. 22:1120 (1954).
Melville, H. W., Trans. Faraday Soc. 32(7):1017 (1936).
Mikusinski, J., Operational Calculus, Pergamon Press, Inc., New York (1956).
Miller, R. C., and P. Kush, J. Chem. Phys. 25:860 (1956).
Milne, T. A., J. Chem. Phys. 28:717 (1958).
Mingwaldt, C. P., Optic 9:248 (1952).
Moon, P., The Scientific Basis of Illuminating Engineering, New York (1936).
Mott, N. F., and H. S. W. Massey, Theory of Atomic Collisions, Oxford University Press, New York (1949) [Russian translation:IL (1951)].
Motzfeld, K., J. Phys. Chem. 59:139 (1955).
Naumov, A. I., Pribory i Tekhn. Eksperim., No. 5:143 (1962); Zh. Tekhn. Fiz. 33(1):127 (1963).
Nesmeyanov, A. N., Vapor Pressures of the Chemical Elements, Izd. Akad. Nauk SSSR (1961).
Nutt, C. W., J. S. M. Botterill, G. Thorpe, and C. W. Penmore, Trans. Faraday Soc. 55:1500 (1959).
Owen, S. P., Phil. Mag. 39:231 359 (1920).
Patterson, G. N., Molecular Flow of Gases, John Wiley & Sons, Inc. New York (1956).
Pazukhin V. A., and A. Ya. Fisher, Vacuum in Metallurgy, GNTI (1956).
Pollard, W. G., and R. D. Present, Phys. Rev. 73:762 (1948).
Polyak, G. L., Zh. Tekhn. Fiz. 3:53 (1935); Zh. Tekhn. Fiz. 5:436 (1935); Izv. Énerget. Inst. im. G. M. Krzhizhanovskogo, Akad. Nauk SSSR, No. 3:53 (1935).
Preter, C. D., Chem. Eng. Sci 8:284 (1958).
Privalov, I. I., Integral Equations, Moscow-Leningrad (1935).
Pugachev, V. S., Theory of Stochastic Functions, Fizmatgiz (1960).
Ramsey, N., Molecular Beams, The Clarendon Press, Oxford (1956).
Rapp, R. A., J. P. Hirth, and G. M. Pound, J. Chem. Phys. 34:181 (1961).

Roberts, J. K., Heat and Thermodynamics, Interscience Publishers, Inc. New York (1960).
Roberts, J. K., Proc. Roy. Soc. A161:141 (1937); Rept. Progr. Phys. 7:303 (1940).
Rodigin, N. M., and É. N. Rodigina, Chemical Chain Reactions (A Mathematical Analysis and Calculation), Izd. Akad. Nauk SSSR (1960).
Rossman, M. G., and J. Yarwood, J. Chem. Phys. 21:1406 (1953); Brit. J. Appl. Phys. 5:7 (1954).
Sapozhnikov, R. A., Theoretical Photometry, GÉI, Moscow-Leningrad (1960).
Schilson, R. E., Ph. D. Thesis, University of Minnesota, Minneapolis, Minn. (1957) (cited by Tinkler and Metzner, 1961, which see).
Searcy, A. W., and R. A. McNees, J. Am. Chem. Soc. 75:1578 (1953).
Sevast'yanov, B. A., Usp. Matem. Nauk 6(6):47 (1951); Teoriya Veroyatnostii i ee Primeneniya 3(2):121 (1958).
Sheftal, N. N., (ed.), Processes of the Growth and Pulling of Single Crystals [Russian translation] IL (1963).
Skobelkin, V. I., and N. I. Yushchenkova, Zh. Tekhn. Fiz. 24(10):1879 (1954).
Smith, K. F., Molecular Beams, John Wiley & Sons, Inc. New York (1955).
Smoluchowski, M, V., Ann. Phys. 33:1559 (1910).
Sparrow, E. M., L. U. Albers, and E. R. G. Eckert, J. Heat Transfer 84:73 (1962).
Speiser, R., and H. L. Johnston, Trans. Am. Soc. Metals 42:283 (1950); J. Am. Chem. Soc. 75:1469 (1953).
Stehn, J. R., (ed.), The Physics of Intermediate Spectrum Reactors, First edition.
Stern, J. H., and N. W. Gregory, J. Phys. Chem. 61:1226 (1957).
Stern, O., Z. Phys. 7:249 (1921); Z. Phys. 41:563 (1927).
Sumpner, W. E., Proc. Phys. Soc. 12:10 (1892).
Surinov, Yu. A., Izv. Akad. Nauk SSSR, Otd. Tekhn. Nauk, No. 7:981 (1948); Dokl. Akad. Nauk SSSR 72:469 (1950); Izv. Akad. Nauk SSSR, Otd. Tekhn. Nauk, No. 4:543 (1950); Izv. Akad. Nauk SSSR, Otd. Tekhn. Nauk, No. 9:1345 (1950); Dodl. Akad. Nauk SSSR 83(2):223 (1952); Izv. Akad. Nauk SSSR, Otd. Tekhn. Nauk, No. 7:992 (1953); Coll.: Power Engineering Problems, Izd. Akad. Nauk SSSR (1959), p. 423.
Taylor, J. B., and I. Langmuir, Phys. Rev. 51:753 (1937).
Taylor, L. J., Phys. Rev. 35:375 (1930).
Thiese, E. W., Ind. Eng. Chem. 31:916 (1939).
Tinkler, J. D., and A. B. Metzner, Ind. Eng. Chem. 53:663 (1961).
Trapnell, B. M. W., Chemisorption, Academic Press Inc., New York (1955) [Russian translation:IL (1958)].
Tricomi, F. G., Integral Equations, Hafner Publishing Co., New York (1961).
Troitskii, V. S., Zh. Éksperim. i Teor. Fiz. 41, 2(8):389 (1961); Zh. Tekhn. Fiz. 32:488 (1962).
Tsien, H. S., J. Aeronaut. Sci. 13:653 (1946).
Tyagunov, G. A., Design Fundamental of Vacuum Systems, Gosènergoizdat (1948).
Vallander, S. V., and A. V. Belova, Vestn. Leningr. Univ., No. 2 (1960); Vestn. Leningr. Univ. No. 7 (1961).
Vallander, S. V., I. A. Egorova, and M. S. Rudalevskaya, Vestn. Leningr. Univ. No. 13 (1964).
Vallander, S. V., (ed.), Aerodynamics of Rarefied Gases, First Collection, Leningrad (1963).
Vallander, S. V., (ed.). Aerodynamics of Rarefied Gases, Second Collection, Leningrad (1965).
Vekshinskii, S. A., A New Method for the Metallographic Study of Alloys, OGIZ, Gostekhizdat (1944).
Villet, R. H., and R. H. Wilhelm, Ind. Eng. Chem. 53:837 (1961).
Volmer, M., Z. Physik 5:31 (1921).
Walsh, J. W. T., Proc. Phys. Soc. 32:59 (1919-1920); Phil. Mag. 75:1 092 (1929); Photometry, London (1953).
Wehner, G. K., Phys. Rev. 114:1270 (1959); J. Appl. Phys. 31:2305 (1960).
Wertenstein, L., Physique, 4:281 (1923).
Wheeler, Advan. Catalysis 3:250 (1951); Catalysis, Vol. 2, Reinhold Publishing Corp., New York (1955), p. 106.
Whitman, C. J., J. Chem. Phys. 20:161 (1952); J. Chem. Phys. 21:1407 (1953).
Wood, R., Phil. Mag. 30:300 (1915); Phil. Mag. 32:364 (1916).
Yarwood, J., High Vacuum, John Wiley & Sons, Inc. New York (1965).
Zel'dovich, Ya. B., Zh. Fiz. Khim. 13(2):163 (1939).

INDEX

D

E

F

G

H

I

K

L

M

N

O

P

Q

R